T0325697

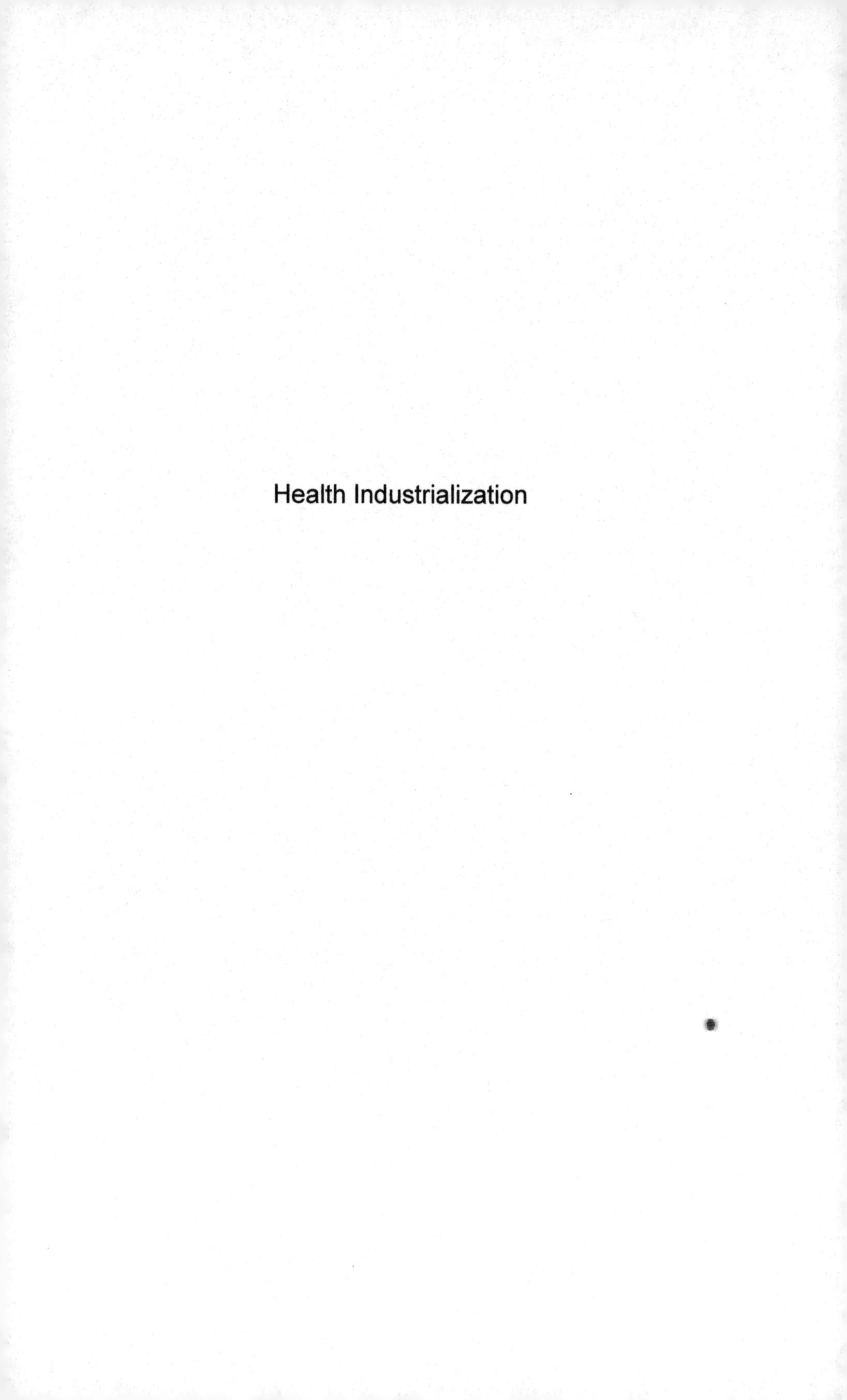

Health Industrialization

Health Industrialization Set

coordinated by
Bruno Salgues

Health Industrialization

Bruno Salgues

First published 2016 in Great Britain and the United States by ISTE Press Ltd and Elsevier Ltd

ISTE Press Ltd
27-37 St George's Road
London SW19 4EU
UK

www.iste.co.uk

Elsevier Ltd
The Boulevard, Langford Lane
Kidlington, Oxford, OX5 1GB
UK

www.elsevier.com

For information on all our publications visit our website at http://store.elsevier.com/

British Library Cataloguing-in-Publication Data
A CIP record for this book is available from the British Library
Library of Congress Cataloging in Publication Data
A catalog record for this book is available from the Library of Congress
ISBN 978-1-78548-147-5

Printed and bound in the UK and US

Contents

Acknowledgments

I would like to thank the following Professors of Medicine for their interesting discussions that have informed this book: Antoine Avignon, Jacques Cinqualbre, Michel Mondain, Eric Renard, as well as André Petitet, emergency physician.

Introduction

Healthcare professionals distinguish between medicine, surgery and diet and lifestyle guidelines. In other words, medicine aims to provide a quantity of life. Men and women would rather remain in good health as long as possible and compensate for the deficiencies that crop up to the best of their abilities. They are looking for quality of life. The result is a tension due to different objectives. This book hypothesizes that this tension is the cause of an industrialization of medicine or health, according to the point of view we choose.

In terms of IT, the machine age started 100 years ago, computers have been around for the past 70 years, the Internet for the past 40 years, the World Wide Web for the past 20 years and more than 5 billion people, out of a total of 7, own a mobile phone. In France, nearly 50% of mobile phones have Internet access. Computer-based Internet access saw a first shift towards mobile access and we are evolving towards a second phase of industrialization that will affect every sector, including the health industry. Each industrial revolution has had repercussions on all sectors of the production system. According to Michel Volle, the advent of computerization has affected nearly every sector – health and education to a lesser extent – but the turning point is now.

Certain authors point out organizational or legal limits. The implementation in all of the infrastructures will require new rights. No one knows what medication someone is taking or if they have it in their bag, but this is no longer the case with IT: everything is recorded and the barcode allows us to detect, at a distance, the presence of the medication. The notion itself of privacy is changing and we need to recreate a new space for it.

Bernard Benhamou talks about a "*droit au silence des puces*" (right to computer chip silence), which would consist of a utopic potential deactivation of systems. Hence, the necessity of modifying the structures of buildings, to this end, will be a feature of the industrialization of health.

The criticisms put forward by other authors are harsher. Jeremy Rifkin outlines the five pillars of "The Third Industrial Revolution", which follows the ones brought about by steam and electricity. These five pillars are related to energy and transportation. As for me, I suggest you go further and understand how this new form of industry is going to spread within the healthcare system to create an actual change in terms of treatment and health. According to Rifkin, "Great *economic transformations in history* occur when *new* communication technology converges with *new energy systems*". These communication technologies and the energy revolution, the storage one in particular, will allow us to outline a kind of health which is more effective but also more movable. This new kind of medicine is then going to overstep the boundaries within which it was formerly confined, especially those of the hospital.

If Jeremy Rifkin sees a transition in the making, other authors deny the significance of IT. Robert Gordon, Robert Solow's disciple, denies the economic effectiveness of computers and introduces the notion of bubble growth: nothing new under the sun according to him. Nicholas Caar states: "IT Doesn't Matter". Jean-Marc Jancovici regards the possible scarcity of energy as inevitable and its decrease as desirable. Thus, information is not one of the weapons of energy saving; it is actually consuming more and more of it.

The definition of health given by the World Health Organization is as follows: "Health is a complete state of physical, mental and social well-being, and not merely the absence of disease or infirmity[1]." To take an interest in health not only involves devoting oneself to the practice of medicine; this is the second aspect of this industrialization of health. Life expectancy and life expectancy in good health are therefore key notions.

---

1 Preamble to the Constitution of the World Health Organization as adopted by the World Health Conference held in New York on 19–22 June 1946. It was signed on 22 July 1946 by the representatives of 61 countries (WHO official acts, no. 2, p. 100) and came into force on 7 April 1948. This definition has not been changed since 1946.

Life expectancy at birth has been used to gauge the effectiveness of medicine for several years. This factor does not take into account the effects of age-related pathologies. The natural degeneration that occurs as we age increasingly seems a disease that we need to cure. The result is a new demand for treatment that has to be reckoned with, since it has to do with the third feature of the industrialization of health: its ability to take care of increasingly older and less autonomous people.

The industrialization of medicine is thus inevitable. This is a phenomenon that has touched all economic sectors apart from medicine and education. ICT has facilitated this industrialization as far as banking and commerce are concerned. The key question then concerns how this industrialization will be implemented. The analysis must be set up in three parts: technologies, societal changes, and the implementation of new institutions and management tools. If the hypothesis about the importance of ICT is true, this work will try to answer the following question: how can we favor the adoption and circulation of e-health or telemedicine systems? We must then find a harmonious kind of development. As it was the case for banking and commerce, this evolution (to avoid saying revolution) will certainly ask similar questions, such as those about digital identification, trust, and power, which in this case becomes "biopower". Biopower is the mode of operation of a power exercised on life: the life of people through their bodies and the life of the population. This notion of biopower was coined by Michel Foucault. This author explains that biopower is gradually replacing the sovereign power, stemming from monarchy, to take someone's life by right, although limited in several countries to life imprisonment. The exercise of this power is what constitutes a government. According to Foucault, before its birth within Nation states, it took root in the "government of souls exercised by the ministers of the Church". This notion of power corresponds not only to questioning the current relationships of domination and balance of power in the health sector, but also to implementing side organizations. The shift from hierarchical power to lateral power will change our way of treating ourselves, but also our way of behaving in society in order to choose healthier lives.

Currently, the significance of biological and medical sciences takes precedence over clinical observation and social reality. Increasingly high-performance analysis methods are spreading. Medical imaging is used to diagnose many diseases. According to some, these are the first signs of industrialization. This is not correct, since every industrialization goes hand in hand with a necessary organization, which neither medical biology nor medical imaging has created.

The new economy emerging from this industrial revolution is taking the exact shape that history predicted. It is therefore distributed and cooperative. It requires a radical reassessment of property relationships and market rules. Therefore, new notions such as identity, access, trust or sharing, which replace the capitalistic vision born from the previous revolution, are becoming predominant; hence the need to reflect on health ethics.[2]

The medical profession cannot be the only one responsible for this development. Doctors are obliged to fight the inevitability of death. This development was not started by international organizations, nations, or administrations responsible for health, since they must take on the role of an institution that vouches for the quality of all technical progress.

Lastly, we have to take into account what we call "the curse of terminology", even when it is appropriate: computing becomes obsolete when faced with digital technology. This is why Michel Volle coined the term "iConomy" to designate the target economy, in which industrialization will be carried out through IT. The industrialization of health will necessarily be started by the patient, before he reaches the point when he finds himself in the hands of medicine. We should no longer use the words "patient" or "ill person" too generally. I propose to replace them with the word *santacteur*[3]. This book will use the definition *santacteur* rather than the rough translation of the American terms "e-patient", "internet patient" or "health seeker".

2 We recommend Jérôme Béranger's book, *Medical Information Systems Ethics* [BER 15].
3 A *santacteur* is a patient who plays an active role within the healthcare system and takes his health into his own hands.

1

Fixed Man, Enhanced Man, Transformed Man

In the past, for a patient, any hospitalization would end in one of three ways: recovery, death or disability, also regarded as "little death". Such was the terrible destiny of a patient. Gradually, chronic diseases have slid into the first two options, thanks to the advent of medical diagnosis. The diagnosis surpassed the symptoms. An example is the definition of diabetes obtained through medical diagnosis: according to international norms, a person suffers from diabetes if his glycaemia is 700 mmol/L (126 mg/ dL) or more after 8 hours of fasting.

1.1. The Anthropocene

Paul J. Crutzen, winner of the Nobel Prize in Chemistry, coined the term "Anthropocene" to mark the epoch that started around 1800, in which men began altering the environment, starting with geology and biology. The changes are attributed to the actions of men upon themselves; hence, the term "Anthropos". One of these is the increase in life expectancy. Jacques Grinevald prefers the term "thermo-industrial society". This visible acceleration of history is at the origin of the new theories that we will tackle later on.

Man has always sought immortality, which remains a mere hypothesis. The industrialization of medicine will be carried out between two extremes:

– ethical medical practice and profit, the dreams of immortality and reality;

– public health policies and demagoguery.

It will also be carried out between:

– sciences and utopias;

– certain methods and sectarian or religious manipulations.

Currently, man-related technologies allow us to carry out three kinds of actions:

– the first is trying to "restore" patients after an accident or an age-related decreased performance. We will call these technologies "restorative technologies";

– the second has to do with an enhancement of performance. These technologies can be of different sorts: artificial intelligence, genetics, nanotechnologies, robotics, etc. It is possible to define, in each of these, a group that we will call "gerontechnology". Other ideas, which seemed to belong to sci-fi a few years ago, have now become a reality. These notions have, therefore, been recently brought up to date;

– finally, new technologies will be able to replace a deficient body, whether it is artificial hearts and pancreases starting to be implanted or exoskeletons that allow humans to walk again.

I prefer this distinction to the one often made between fixed man and enhanced man.

These new medical technologies follow on from the post-Pasteurian[1] trilogy based on the "hygiene-vaccines-antibiotics" triangle, which came to an end in the 1970s with the explosion of medical diagnosis. Between Pasteur's discovery and 1970, i.e. in less than 80 years, average life expectancy increased by 30 years. Hygiene was based on communication and education, and involved several myths. These must also be analyzed through communicational models in relation to health technologies. Philippe Breton, researcher at the CNRS, clearly highlighted the existence of these myths through his critique of Norbert Weiner's philosophy [BRE 99].

1 We refer to Louis Pasteur, who implemented vaccination.

The theory of the four humors was the keystone of medical philosophy for two centuries. According to this school of thought, it is first of all the blood produced by the liver and received by the heart that creates a sanguine, jovial and warm temperament. Then there is phlegm, also called lymph. Attached to the brain, it creates a phlegmatic temperament. The third humor is the yellow bile, also produced by the liver, which characterizes a choleric temperament, inclined to violence. The choleric temperament is said to give an impression of force and control. The last one is black bile, produced by the spleen, which is an expression of anxious or melancholic temperaments. These four humors correspond to the four elements: fire, air, earth, and water. These are ideas based on beliefs and ideologies which rely very little on scientific methods. So, nosography is the science that allows us to classify symptoms and pathologies. Doctors could avail themselves of "nosology" to treat.

For years, after the adoption of Pasteurian medicine, medicine was considered as a fight against agents external to humans. This point of view will only belong to a negligible part of the health field.

"The theme of a communication-centric society unknowingly takes on the responsibility for the utopic ideal of social change that had begun to express itself a century before. This idea of a communication utopia is therefore questionably positioned in the continuation of theories of political change which, in the 19th century, did not postulate either social exclusion or the reinforcement of the role played by the State, and which had started looking for other forms of social regulation [...] that pivoted around technical progress and a new relation to machines, but also around another definition of Man and of the social bond" [BRE 97, p. 62]. This communication-centric society took off with diagnostic tools, such as radiography, born at the end of the 19th Century, and then chemistry, which makes blood tests possible. This technical progress and the notion of the external enemy helps us to forget that life takes each of us along the path of aging and death.

Abdel Omran [OMR 71] has defined three periods in medicine: one characterized by extremely deadly plagues [WER 99] and famines, one characterized by the drop in pandemics – when the "hygiene-vaccines-antibiotics" triangle was effective – and finally the new period characterized by degenerative or "societal" diseases. As a result, it would be possible to create a new man. These new diseases no longer have a single external cause listed by nosology. This way of diagnosing comes to an end at the moment when thousands of new quantifiable forms of diagnosing appear. They no longer correspond to symptoms, but to hypothetical or future symptoms. Together with the appearance of these kinds of diagnoses, some technologies enable us to act directly on the different disabilities that human beings may be led to face. Hence, the birth of a new man: enhanced, android, bionic and even cyborg or hybrid.

1.2. A new man in the face of progress

The notion of progress is not self-evident. It is quite recent. As for the future, it is a process that establishes itself deceptively through keywords that we will group under the term "new man". So, a new man appears through the introduction of technologies performed on man that are described as follows. I have chosen to list these concepts alphabetically.

1.2.1. *Actroid (clones)*

The notion of cloning, so that we could have at our disposal a perfect double that can restart our life, had its hour of glory with the success of animal cloning. Currently, research is focusing more on Actroids, which are artificial "doubles" programmed on someone's appearance, expression, movement and even voice. Roboticists regard them as a solution for helping aged or disabled people.

1.2.2. *Android*

By Android, we mean anything with a human shape. The etymological definition includes anything that may look like a human. The use of this word presupposes that we are not referring to an actual human being.

1.2.3. *Bionic*

Bionic is a portmanteau formed from "biological" and "electronic". Bionics is the branch of science that investigates how a living being receives and processes signals to act. Its goal is to replicate this mechanism in machines and robots.

1.2.4. *Cyborg*

The term "Cyborg" designates a living organism made of organic matter and cybernetic elements (electrical circuits, mechanical systems). Cyborg is a portmanteau created from combining the words "cybernetic" and "organism".

1.2.5. *Enhanced man*

This concept defines those technologies that try to enhance human abilities beyond their natural biological evolution. In this context, the steps taken in terms of fixing, transforming and enhancing are notable. In contrast with this definition, there are also some rather old concepts, which are mainly the product of "sci-fi".

1.2.6. *Hybrid man*

The term hybrid man designates a new type of man, a mixture of two bodies combined into one. One of them is natural; the other is a product of technology. Experts are analyzing a point of reference that can vary from simple juxtaposition to perfect fusion. This term comes from classical Latin "ibrida", which means "bastard, mongrel". This implies, by analogy with agriculture, that the hybrid man cannot reproduce as such. He will only give birth to "simple" men, i.e. beings that do not possess all human qualities and functions.

The robot-based approach completely omits the evolutionary component of man that necessarily leads to death. Despite everything, the "robot" model remains the most prominent. On the other hand, "the immense set of degenerative diseases is hard to distinguish from the natural effects of aging, even if the age at which they develop varies in relation to the person" [PÉR 12, p. 51]. Deafness, cataracts, and mobility issues affect nearly everyone at a certain age. Prostheses have become a solution; hence, new technologies have been adapted to this situation.

1.2.7. *Hybris or hubris*

This is a very old perspective that dates back to the Greeks. It is a form of immoderate pretentiousness, an extreme kind of pride or arrogance made possible by the possession of a piece of technology. This excessive pretentiousness often involves losing touch with reality. It is embodied in a form of overestimation of one's own competence or ability, particularly when the person displaying it is in a position of power. Certain types of technological equipment such as limb prostheses or pacemakers can now put men in positions of this kind. The example of the Paralympic runner who wanted to compete against Olympic runners illustrates this point.

"Google Glass" – glasses that allow us to simultaneously record and display images right before our eyes – is part of these "hybristic" developments.

This list is mentioned merely to create a debate, since we do not think that the above-mentioned forms of humans will develop. On the other hand, the bet on human evolution as enabled by technology is certainly going to be won.

1.3. Fundamental technologies

These technologies are mainly used for aged or non-autonomous people.

1.3.1. *Gerontechnology*

Many geriatricians are convinced that technology can compensate those deficiencies that affect people over a lifetime. When aging, our sight as well as our hearing decrease, but diseases and accidents can also lead to disabilities. These kinds of technologies are called gerontechnology. The two questions that crop up in relation to the development of these technologies are the following:

– how to turn the process of aging into a potential source of economic profit?

– what risks do they involve?

The development of these technologies is most likely to go separate ways and to require the presence of several actors with different profiles and competences. As is the case for drug therapy, useful technologies rely on whether or not they bring actual benefits.

Providers have to show that technologies help humans. However, the mere fact that it is not (and never will be) social security that buys tablets and mobile phones, but the people affected themselves, raises the problem of the quality of this offer.

1.3.2. *The robolution*

Robolution is a term coined by Bruno Bonell, who thinks that rather than universal robots, there will be a series of robots and intelligent units able to facilitate household life.

The term "robolution" is proposed to designate a series of robots and intelligent units. Among these, we should mention exoskeletons, which help carrying heavy weights or hemiplegic people, an interactive therapeutic baby seal for children or patients suffering from Alzheimer's disease, and service robots for disabled people. Following this logic, the Aldebaran Robotics society came up with an android robot, Nao, that has sold several thousand units since 2008. This kind of robot can tell stories, answer questions, etc. This approach goes hand in hand with the Internet of Things, while also remaining a potential competitor.

The impact of this robolution has been strengthened by the emergence of 3D printers. So, it is possible to fabricate prosthetic arms adaptable to each individual starting from a public and free project. As proof of how patients are involved in the advent of the industrialization of health, an association called E-Nable was set up to link people who have lost limbs with users of 3D printers able to customize the project.[2] "Maxence", a 6-year-old from Isère, was one of the first French people to benefit from this approach.

1.3.3. *Anthropotechnics*

"Anthropotechnics appears to be a many-sided service of human biological transformation for purposes including performance, identity quest, freedom, standardization, etc.".

---

2 The website of this association is http://enablingthefuture.org/.

This definition, proposed by Jérôme Goffette, refers to a list of technologies and approaches: aesthetic transformations, physical doping, mental doping, humor modulation, control over sexuality, cyborgization, atypical assisted reproductions, mechanical ventilation, walking assistance, implants to see, listen, or hear, etc.

1.3.4. *Nanorobotics or fog utility technology*

The nanobots proposed by Kurzweil are microbots, the first of which are emerging from their experimental stage. They are swallowable capsules that examine the digestive system. However, this term can also refer to robots that roam all over our blood system and cleanse our blood or repair tissues.

The foglets are nanobots that can lie side by side and create objects useful for humans.

1.3.5. *Diagnostic technology*

Diagnostic technologies saw their hour of glory with the advent of chronic diseases. It is the diagnosis that makes the diseases chronic, and some of them are asymptomatic. They have no single and identifiable external cause. They include cancer, cardiovascular diseases, neurodegenerative diseases such as Alzheimer's and Parkinson's and, more recently, bone diseases.

The DMS (Diagnostic Medical System) society, based in Montpellier, has developed a DXA osteodensitometry with 3D reconstruction, which will allow us to diagnose osteoporosis, a disease that leads to fractures for one woman in three and one man in five from their 50s onwards. These tools have often provided diagnoses that the medical staff did not know how to handle.

Most often, it is, therefore, the diagnosis that makes the disease chronic.

Chlamydia is one of these diseases. In Ireland, for instance, a study showed that 80% of women and 50% of men infected with Chlamydia show no symptoms. Chlamydia is an infection caused by the Chlamydia trachomatis bacterium. It even seems that we can be infected more than once in our lifetime. It is an asymptomatic disease. People get infected without realizing it, since the disease has no symptoms, and can, therefore, in turn transmit the disease.

If symptoms are detected, they appear 2 to 5 weeks after the transmission. They are as follows: vaginal discharges or bleeding after sexual intercourse and between vaginal menstruations, discharge from the penis or the anus, stinging or burning when urinating, pains in the lower abdomen or during intercourse. The advice is to see a GP.

Chlamydia, when untreated, can have dire consequences on women's health: infertility, ectopic pregnancy (in the Fallopian tubes), and chronic pains in the lower abdomen. In men, it can lead to chronic prostate infection (prostatitis), one strain of which is caused by this bacterium. It also increases the risk of catching or transmitting HIV, as well as the risk of developing cancer.

There is no vaccine against chlamydia and only "sexual protection" remains effective against it. Condoms are, therefore, the best kind of protection. Pfizer's Azithromycin can treat chlamydia effectively.

This bacterium is sexually transmitted but it can also spread "orally", although in this case the risk of transmission is lower.

Box 1.1. *Chlamydia, a diagnostic kind of disease*

Sarcopenia (a weakening of the muscles), described in 1989 by Irwen Rosenberg, is another example of the impact of medical diagnosis. It inevitably leads to death but it represents the great progress made by medical diagnosis, since the diagnosis allows us to follow its development.

1.3.6. *Genetics*

Genetics is certainly the kind of technology that will pave the way for the industrialization of health. Genetic diseases are linked to the malfunctioning of a gene and to other factors that are more or less known. If they are hereditary diseases, the transmission takes place whenever a part of the parent's genome is transmitted to the infant during sexual reproduction.

However, external factors such as radioactivity or chemicals can lead to the same situation. A genetic disease can be described as an anomaly or, in scientific terms, the mutation of a gene or of an allele. This situation will result in the production of an abnormal protein. This protein can lead to different consequences in relation to diseases that can appear at birth or later on, it can lead to the development of cancer, etc. Genetics on its own has shown its limitations and researchers have joined it with epigenetics.

Organizations such as Orphanet[3], the portal for rare diseases and orphan drugs, or the *Online Mendelian Inheritance in Man* (OMIM) of Baltimore's Johns Hopkins University, catalog genetic diseases and the genes associated with them. This list is more or less updated on important medical portals and even on Wikipedia.

In 2012, experts estimated the number of known genetic diseases at 10,000, and reportedly discover five new ones each week.

1.4. Debates on technologies and men

These debates have appeared in response to certain movements promoting new kind of man greatly enhanced by technologies. The creation of such a paradigm omits, and sometimes effaces, the natural evolution that leads to death.

1.4.1. *Transhumanism*

Transhumanism is a philosophical doctrine that defines itself as a cultural and intellectual movement. It analyzes, advocates and encourages the use of certain technologies to enhance the human condition, both physically and mentally, beyond the constraints of natural biological evolution.

Transhumanism is governed by an international association, the WTA (World Transhumanist Association), founded in 1998 by philosophers Nick Bostrom and David Pearce. Transhumanism has branches in numerous countries. The association called >H has recently adopted the abbreviated form H+. The former website of the association suddenly shut down in 2009. Currently it is possible to find information on the "humanity plus" website. It

3 Its website is http://www.orpha.net.

is very interesting to note that several countries have developed research centers and even founded universities in relation to this topic, whereas others have dismissed these approaches.

1.4.2. *Replacement anthropotechnics*

These new notions of man lead researchers to question the boundary between replacement and enhancement; whether because the idea itself of transformation is out of the question, or because it is considered unethical, or again because it is seen as something that can potentially be reduced to either possibility (replacement or enhancement).

According to philosopher Jérôme Goffette, current medicine is a form of anthropotechnics that attempts to meet the two needs of replacement and enhancement. Therefore, this difference would not exist for different reasons. He goes even further and writes that "several contemporary 'medical' practices have dismissed treatment and the fight against diseases in order to 'enhance' man". Let us point out, however, that doctors distinguish between the two approaches in their practices. In spite of everything, the philosopher sees two breaking points taking shape: the first in relation to the cry for help against death and suffering, and the second in relation to the medical duty to assist. Second argument: we should then have a clear notion of normality. Finally, it is impossible to make a precise distinction between replacement and enhancement technologies and, besides, over the course of history, there has always been a shift from replacement technology towards enhancement technology.

The debate about this new man refers to the notion of "assisted man", which involves the risk that people may no longer be able to carry out the tasks they are expected to.

1.4.3. *Algeny*

Algeny is a term used and proposed by Jeremy Rifkin, who sees a new form of humanity in these new technologies: a sort of biological plasticity that might lead mankind towards a new era.

1.5. World, *mondialisation*[4] and health

If the idea of industrialization is often linked to that of an assembly-line work, we have to point out that this last element has gained all of its strength in the process of *mondialisation*. Before tackling this latter development, we deem it necessary to think about the meaning and use of the word "world". It is actually a catch-all term, with the potential to hide our contradictions as well as our ambiguities. It becomes important to analyze its framework in relation to the health industry.

1.5.1. *A tool for measuring speed*

The first connotation of the term – in the sense "around the world" – turns it into a tool for measuring distance and speed, as such books as Jules Verne's *Around the World in Eighty Days* reveal. This inevitably leads us to the notion of rapidity of information diffusion. Thanks to networks, a piece of information can be propagated around the world in record time.

It is, therefore, possible, at least at a first stage, to conceive a *mondialisation* of the offer of treatment that allows patients to be treated more promptly.

A second approach consists of the dynamic view of human beings and dispositions, characterized by unstable equilibria and elements of the dynamics of diffusion. The asymptomatic carrier defined by Robert Koch illustrates this point. Microorganisms are inside the human being, but they are not active yet.

1.5.2. *A better world*

The term "world" is also used to mean "the other world", a better world. The myths about IT according to which – as Jacques Perriault underlines – there is a positive *mondialisation* certainly rely on this statement. Shouldn't modern medicine lead to a better world? In this case, epistemology and the history of health have described a healthy being that is simultaneously an

4 The differences between *mondialisation*, *mondialité* and *globalization* will be pointed out in the next few sections, where a definition for each of these terms will also be provided. *Mondialisation* and *mondialité* have not been translated since English lacks any suitable equivalent able to convey the same meaning.

ideal and a purpose of medicine. This ontological approach is characterized by an increase in the size of the ontologies. Therefore, the same biochemical diagnosis – the amount of sugar – has led to the description of two very different diseases: type 1 and type 2 diabetes. This ontological evolution also presents itself in terms of multiplication. Therefore, there are several hundred medical ontologies with specific uses.

1.5.3. *A world outside humans*

The world is defined as "that which is not part of oneself", which consequently corresponds to everyone else. To go into town means to go out into the world; this is how the world defines simultaneously wholeness and difference. The world is, therefore, considered a boundary as well as a frontier. It should tend to be widening with the development of technological progress. This is the theory developed by François Perroux: "Science, technology, and industry widen the circle of contacts and the circle of communications between beings and their groups" [PER 60, p. 68]. We should not forget that our modern technologies will change the world. Just think of what would become of wheelchair access policies if exoskeletons, intelligent prostheses and robots were to become widespread. This domination of "the automated man" is characterized by mechanical therapeutic actions. Mechanics, together with electronics, acts to fight diseases or disabilities or to eliminate their effects.

1.6. *Mondialisation*, globalization and *mondialité*

Globalization and *mondialité* are the two main words used in relation to the dynamics of industrialization.

Mondialisation versus globalization

According to Marc Augé:

> "The term *mondialisation* refers to two kinds of realities: on one hand, what we call globalization, which corresponds to the spreading of the so-called liberal market and of ICT networks all over the world; on the other hand, what we may call planetary awareness or planetarization, which itself consists of two aspects".

Studying modern medicine certainly involves the study of these technological networks. They are even the most tangible representations of this *mondialisation*. Planetary awareness, in this case, is twofold. The first component can be qualified as ecological:

> "Every day we are aware of living on the same fragile and threatened planet, infinitely small in an infinitely large universe".

Health conditions vary from one geographical area to another, but medical technologies are spreading globally and becoming reliant on technologies, if only on testing technologies and medical imaging. Therefore, this allows a Frenchman or a German to have their teeth fixed in Hungary. Thus, we will have to imagine the breakdown of these technological tools and find a solution to this situation. The second form of awareness is social:

> "We are also aware of the ever-growing gap between the richest and the poorest, and of the parallel gap between knowledge and ignorance".

What to think of this planetary ecological or social awareness that is never put into action? The difference in the price of medications, even despite industrial production, illustrates this point. The result is that:

> "The term 'globalization' refers to the existence of a global liberal market – or deemed such – and of a worldwide technological network, to which, however, a large number of people have no access yet. The global world is then a network-centric world, a system defined by spatial and also economic, technological, and political parameters" [AUG 08, p. 41].

The key factor of this globalization is the necessary conformity to the least expensive or best-performing technical solutions. These have the advantage of allowing access to healthcare in the poorest countries. Still, industrialists have to agree to lower the prices of medications and supply for low-income countries.

1.6.1. *The notion of mondialité*

In relation to this conjunction of globalization and *mondialisation*, Philippe Zarifian chooses the term *mondialité*. "*Mondialité* can be defined straight away and quite simply as the human sense of belonging to the same world if, by 'world', we mean the planet Earth". He finds it strange that this sense of belonging is seldom mentioned by humans [ZAR 04, p. 7].

This notion of belonging to a world, which Zarifian at first qualifies as the entirety of our planet, introduces a new problem, if we take into account the presence of "sub-worlds" or of the world communities to which an individual belongs. Why shouldn't a man or a woman from a certain country benefit from the medical progress carried out by another? Will those who are excluded from healthcare systems be left out on a merely financial basis, or will industrialization lower the costs to such an extent that all forms of medicine will be available to everyone? This approach leads us to believe that, with the advent of this industrialization, the development of a form of medical tourism is inevitable. The *santacteur* we previously mentioned will find out where that is taking place and travel there.

Another medical approach to this concept has to do with the global diffusion of viruses and bacteria like H1N1 influenza or Severe Acute Respiratory Syndrome (SARS). SARS is an infectious pulmonary disease (pneumonia) caused by the SARS-CoV[5] virus, which literally took off with one of the sources in Hong Kong to infect people from France to Canada. On the other hand, a recent and geographically closer virus, the MERS-CoV[6], has spread to a lesser extent in the Persian Gulf countries ever since its appearance.

1.7. Globalization, internationalization, localization

At this stage, it seems important to focus on the case of technicians and engineers. Software building experts propose three approaches.

1.7.1. *Some additional definitions*

Globalization consists of creating a "global experience" of software, which, therefore, becomes the point of reference. This process has been widely used by companies such as Microsoft. It is often based on education and information diffused effectively, discretely, indirectly and targeted in relation to the use of the product.

5 Severe Acute Respiratory Syndrome Corona Virus.
6 Middle East Respiratory Syndrome Corona Virus.

In the medical field, globalization was essentially controlled by the WHO, which diffused information about risks. The industrialization of health will definitely lead to the appearance of companies acting this way.

Internationalization consists of creating a framework which is as neutral as possible and adapting it to each local situation. This approach has been used for the websites of multinational companies.

Localization consists in building an interface with versions that can be adapted to different cultures. Educational software publishers have focused on this kind of approach. Since each country has its own specific notions of education, software can seem to vary widely from one area to the other but it is actually built on the same foundations.

The problem consists in understanding the kind of shape that the industrialized medicine of the future will take. We will not consider this debate any longer, which we would not consider too productive given the current developments in terms of health. Let us remark right away that such systems as radiography, CT scans, and MRI are related to globalization.

1.7.2. *Types of mondialisation*

What does ICT contribute when faced with the pressure of *mondialisation*? Two answers seem possible: the destructive creation that begets a new form of more industrial society and, on the other hand, a compulsory connection to networks.

At the time of Jean Voge [VOG 63] and Abraham Moles [MOL 71], cycles constituted the only model. Economists were discussing Kondratiev waves (1992) of destructive creation. Is *mondialisation* the enemy to fight, the danger or the threat that challenges this model? The answer can only be affirmative when we consider the list of works published on this topic. On the other hand, no sign or reading leads us to think that the object of our study generates a kind of particular *mondialisation* that has the potential to create a new society. This approach will constitute the basis for our conclusion. We will try to determine the features of this society in which medicine will be more industrial.

Logically, we should pay attention to the other kinds of cycles defined by economists, such as the Kitchin cycle (3–4 years), the Juglar cycle (8–10 years), the Kuznets cycle (15–25 years) and the Kondratiev wave (40–60 years). It is possible to claim that the numerous experiments conducted in terms of telemedicine and e-health fall within a Kitchin cycle and consequently do not lead to lasting solutions. The Juglar cycle is now the renewal cycle of large healthcare equipment. We should point out that this has become 7 years shorter, which corresponds to the period needed to pay off the heavy equipment necessary for radiography, CT scans or MRI.

With MRI and the explosion of diagnostic methods, we can see, as far as our topic is concerned, the emergence of a new Kondratiev wave. It is the end of the "penicillin era". The problem we face at this stage consists of the link between the era of degenerative diseases, determined by Abdel Omran, and the rise of diagnostic technologies.

1.7.3. *Technosciences*

Alain Gras [GRA 97] considers *mondialisation* as an obligation to get connected. Technosciences constantly require the individual to get connected to an increasing number of technological networks, under which other, kinds of networks sometimes lie hidden. The individual is incorporated in infrastructures, which, however, have a flaw. Technical choices are made in mysterious and often unknown places. Sometimes it is a matter of safeguarding the structure of the project. In other cases, we have to do with a form of power obsession stemming from certain people. Finally, we have to do with new forms of power: a power upon the body that we will call "biopower". The individual is condemned to connect to the network, otherwise he will be condemned [GRA 97]. In other words, in order to benefit from medical progress, connection will become essential to data results of every kind, which come from publications as numerous as they are contradictory.

1.7.4. *Evidence-based medicine*

Current medical teaching uses the concept of Evidence Based Medicine (EBM), namely medicine based on proofs. These proofs are certainly a product of the *mondialisation* of medicine and are published in renowned journals. This approach requires four points of verification for any kind of medical action: the existence of experimental data, the building of theoretical support, a statistical proof and the presence of an authority that validates all of these.

EBM, as it is being practiced now, has four consequences. The first is the confusion between risk factors, data and the disease, which is certainly the most insidious and veiled. The second consists in a rapid shift from the correlations found through studies to actual causal links. The result is that medicine will treat patients who are not ill and, conversely, that doctors can overlook patients in need. Finally, the success of a drug-based and technical solution is favored. This can potentially entail overmedication, overtreatment and unnecessary operations, all actions that healthcare professionals perform in good faith. Lastly, this approach overestimates figures and underestimates clinical observation.

The industrialization of health will have to involve EBM, but under no circumstances will this be the only basis it will rely on.

1.7.5. *The shortage system*

Medicine often works in "shortage management" mode, which is at the core of technical progress. *Mondialisation* has added a new dimension.

The possibility of performing transplants to treat patients has to be faced with the limited number of donors. It has given rise to the development of new medications able to widen the field of possible transplants. These same limits drive the development of artificial organs: initially external machines such as dialysis units, with the hope of turning them into intra-corporal devices, like artificial hearts.

The shift from a mundialized production to local needs causes problems. The *mondialisation* of pharmaceutical production is an example we may use as a starting point to understand the risks entailed within the process of *mondialisation*.

The report we have drawn up is standard in the industrial field, but raises problems in the health industry, where we must prevent patients from lacking the availability of medication. We have to note that presently around 60–80% of all active substances destined for the dispensaries of pharmacies pass through the hands of wholesalers and distributors. As for shortages, the figures announced by the Academy of Pharmacy are as follows. Every day there is a 5% shortage of all medications ordered, and 50% of the shortages last more than 4 days. The table below is in line with the position of the Academy of Pharmacy [ACA 13].

Type of shortage	*Solutions*
Situations of shortage or discontinuation in the production of certain active substances still useful in terms of public health	To draw a list of the active substances that have become of public domain, the shortage of which could lead to public health problems To make European authorities regularly inventory shortage risks and relocation needs
Shortages due to the lack of quality of imported raw substances	To set up a European index of these substances and to make it public to those authorized to launch them on the market To exchange on a European level, the results of the inspections conducted in non-member countries, and to find ways to carry them out
Drug shortages due to the discontinuation in the production of certain low-return pharmaceutical substances	To implement a way of notifying the authorities To revalue prices in order to favor the production To organize a way of transferring the production on to a public pharmaceutical organization
Drug stock shortages linked to production-related quality flaws and to the policy conducted in terms of production – especially of certain kinds of medications – or industrial stock management	To shift from an information-centric culture to a dialogue-based one To introduce risk assessment methods To guarantee supply by bettering stock and production management
Supply shortages linked to the distribution network	To reconsider the pertinence of stock management rules To monitor the distributors' re-exports

Difficulties related to the public tender system (hospitals)	To reflect on public powers in order to avoid the appearance of excessive invitations to tender To integrate clauses concerning total quality management and supply security into the criteria regulating invitations to tender
Shortage of specific drugs (orphan, pediatric, radiopharmaceutical drugs)	To think specifically about the management of orphan diseases that require active ingredients of pharmaceutical standard and to encourage the European production of these ingredients To provide hospital pharmacies with raw substances of pharmaceutical standards To start thinking about the future of the production of radioisotopes for medical purposes
Specific pharmacy-related difficulties	To implement access to information concerning shortages and availability To set up exchange systems

Table 1.1. *Causes of shortages (medications)*

1.7.6. *Life Meccano*

This approach was suggested in the 1990s. It is mainly known thanks to François Jacob's speech delivered for his reception at the French Academy.

> "All the creatures that inhabit this earth, whatever their environment, size, or means of subsistence – snail, lobster, fly, giraffe or human being – all turn out to be made from molecules that are more or less identical.
>
> And likewise, from yeast to humans, there are groups of closely related molecules that serve to assure universal life functions, such as cell division or signaling between the cell membrane and the nucleus (…). It appears, then, that all life

> forms are constructed with the same modules, distributed in different ways".

This sentence propels science into the search for these life "atoms". It would therefore be sufficient to link diseases to the presence of this or that element.

"The living world is a sort of combinatorial system composed of a finite number of parts, like the product of a gigantic Meccano set; it is the result of a ceaseless process of evolutionary tinkering." Here lie the foundations of a whole theory of "human programming", based on the same principles of computer programming.

Nowadays, this mechanistic notion of medicine seems too simplistic.

2

The Necessary Industrialization of Medicine

The industrialization of medicine has become necessary since the non-industrial use of technology is costly and can no longer support systems. Auguste Comte[1], Emile Durkheim[2] and Ferdinand Tonnies[3] all agree on the definition of industrialization. Auguste Comte put forward the “Law of Three Stages”. According to him, the human mind consecutively goes through “the theological stage”, or imaginary, “the metaphysical stage”, or abstract, and finally “the positive age”, in which the only truth is that which is attained exclusively through science. According to Emile Durkheim, industrialization and its division of labor constitutes the only peaceful solution in terms of coexistence in current societies, although he acknowledges pathological forms of this division. According to Tonnies, social peace depends on the state of equilibrium reached between human relations and rational personal interest, which includes an industrialization of actions. This corresponds to two kinds of human will: organic will (pleasure, habits and memory) and rational will (reflection and decision).

DEFINITION 2.1.– *Industrialization is the process of transformation of the production system of a society through technology.*

This economic definition allows us to define the concept of industrial revolution and to introduce the notion of division of labor. The problem that

1 This theory is described by Auguste Comte [COM 98].
2 The work used for Durkehim [DUR 07].
3 François Jacob’s acceptance speech made at the French Academy on 20 November 1997. The reference work for Tonnies [TON 87].

arises at this stage, which we will deal with later on, consists of asking ourselves how medicine and its current actors will relate to this new process of industrialization.

It would be possible to reflect, as Emile Durkheim did, on how this industrialization has contributed to the isolation of the individual. This individualization actually makes the healthcare process pivot around the individual. It is an interesting paradox: a system set up to treat individuals that, however, alienates them. It may be the advent of customized medicine.

On the contrary, it is possible to question the economic impact and the revenues-expenditures equilibrium, in terms of health, of this industrialization. It is then that the problem of the economic model of health arises.

2.1. Medical innovation as a factor of industrialization

Innovation constitutes the basis for industrialization. Therefore, it becomes interesting to address the topic of health innovation. It is a major aspect of the evolution of our societies, "mirroring its ideologies and aspirations". Innovation is defined as "a process of recombination of existing knowledge, be it implied or formalized, empirical or codified, technical or scientific" [CAR 11, p. 9].

2.1.1. *Process innovation and product innovation*

We often make a distinction between process and product innovation. In the former case, innovation means performing a twofold action: raising and then solving a problem. In the latter, innovation is an answer to a need of the market or society. Modern medicine has been based on these two kinds of innovation for a long time.

Medical innovation started with product innovation. Biotechnologies were initially implemented by pharmaceutical industries. Today, 30% to 40% of the new molecules indexed by the health authorities in the United States stem from biotechnologies, whereas the figure is slightly lower in Europe. These molecules represent 50% of the portfolio of the large pharmaceutical laboratories that have raced to purchase start-ups in the field. They are pharmaceutical products no longer produced chemically, but created from bacteria, yeast or animal, and soon human, cells.

After being constrained by a demand that derived from the healthcare system, the patient, namely the main beneficiary, is more and more involved in the changes brought about by innovation.

What characterizes recent progress is not the pursuit of these developments. Process innovation merely aims to reduce the visible and quantifiable malfunctions of the healthcare system. Medical science consists then in enhancing and sometimes creating treatment protocols. Product innovation consists in meeting the demand coming from doctors, as is the case for medical robotics, or patients, for example for motorized wheelchairs fitted with robotic arms. Currently, the most significant development comes from the patient's demand in relation to his place within the global process, hence the creation of technical devices that combine products and processes.

2.1.2. *The patient's role*

The current situation is dominated by the patients' claims to a more prominent role within the health ecosystem. Progress is dominated by patients laying claims to technology and knowledge to a greater extent. In addition, patients are coming together in associations and institutions. If the medical profession at first attempted to regain ground, these modern organizations are excluding medical power little by little without discarding its competences. The Kouchner Law and then the French "Hospital, patients, health, territories" reform act have provided the patient with more legal power. This is how new care devices are born.

Philosophers, historians and sociologists interested in medicine all notice the growing gap between science, medical techniques and health, the latter being necessarily personalized. It is only recently that the notion of "quality of life" has taken root in the medical field. Luc Périno dates this development back to the 1980s [PÉR 12, p. 51].

2.1.3. *Personalized medicine*

Personalized medicine has thus become a trend. It is not a matter of giving a patient a unique solution. We have to become aware that patients can respond to a treatment to a greater or lesser degree. Personalized medicine wants to be able to find out why. This choice of treating or not treating a patient in a certain way is made in relation to biological markers, some of which are merely genetic.

For some of these diseases, the progress made has been very significant. Thus, an American consortium announces to have identified almost all the mutations involved in acute myeloid leukemia (AML) in adults. Since it is one of the hardest kinds of cancers to treat, this new piece of knowledge constitutes a fast track to the identification of therapies and markers. According to these researchers, it should also enable us to predict with more precision the seriousness of the disease in each patient.

The progress being currently made consists in developing medications and treatments in relation to the biological target, which must therefore be detected. This entails a very significant revitalization of diagnostic procedures.

2.1.4. *Towards a new world*

This new industrial world is characterized by some constants, but it comes with new dangers.

The main constants are:

– the automatization of repetitive tasks, which started out in factories and has affected, for example, novelty searches carried out by lawyers. In the health field, this process will certainly be applied to medical diagnosis;

– the nature of products is altered since the goods sold are a mixture of goods and services, and the service is the quality element. For elevators as much as copy machines, quality depends on the maintenance service and, by analogy, it is a safe bet that the patient's follow-up care after he is discharged from the hospital will be a key point;

– interconnected goods and services are linked to partnership networks and the interoperability of the partnership is ensured by information relationships. These are more or less fixed within an information system. It is necessary to define the products, the production nodes and the factories allocated close to the clients. In the health industry, this will lead to a redefinition of healthcare sites, which have concentrated in large urban areas over the past few years;

– it is engineering that provides employment and, as for services, the workforce is replaced by mental force. It is necessary to know how to take decisions and solve problems. The result will be a new approach in the training of doctors, certainly less practical, and the advent of new professions.

2.2. The notion of apparatus

We are interested in the notion of apparatus, since it seems essential to this debate as it is described in Giorgio Agamben's work on devices. In relation to our analysis, the author observes that this notion of apparatus is a stronger concept than that of Machine for Communicating, advanced by Pierre Schaeffer. MRI and CT scans are not only machines for communicating images of the patient, but become part of the analysis, and sometimes for the measurement of the tools employed on a patient. This notion of apparatus also allows us to leave behind the concept of diagnostic machines.

2.2.1. *Agamben's apparatus*

This notion of an apparatus taken from Agamben is quite close to the concept of a technical system used by Jacques Perriault in his work on the logic of uses. "The concept of a technical system is taken from Bertrand Gille himself, who defined it as a set of converging techniques whose combination contributes to the creation of a well-defined technical act" [PER 08, p. 26]. We prefer the notion of apparatus since this term evokes, in addition to a sum of technologies, the notions of function and application which we will use later on.

Agamben, in his book *What is an Apparatus?*, studied the relationship between our lives and apparatuses. He also wondered how we are supposed to act when faced with this situation. Apparatuses are systems that, by now, contain our existence. From tools only available to healthcare professionals, they will become accessible to the consumer. Consequently, they will be soon seen by men as mere consumer products. A perfect example of this development is the assessment system used for people suffering from diabetes, which allows for communication via mobile phones.

Apparatuses change our personalities. If we refer to Agamben, then the question becomes: which strategy should we adopt in our daily close engagement with apparatuses? This relation gives birth to new societal forms that we will be led to study later on. Biologists' laboratories will be the first ones to be affected by this development. Analysis machines being more and more effective and inexpensive, this leads to a concentration of analysis centers, and portable apparatuses which enable patient self-sampling.

The development of these systems of biological analysis clearly shows that we are entering a new Kondratiev cycle in which reading the color of a strip after a reaction does indeed belong in the past.

Development	Year
First lab control mechanism and results computerized on a database	1970
Real-time results in the operating room	1970
Command-line interface of measurement instruments	1970
System of lab information accessible interactively in bed or in the operating room	1985
Integration of lab messages into other systems	1995
Systems of auto-verification of the results provided by machines	2000
Normalized interfaces of analysis systems based on the Internet and the Web	2000
Possibility of collecting patient data remotely	2000
Possibility of reporting multi-channel results (letter, Web, fax, mobile technology such as texting)	2005
Possibility of collecting data directly from the patient via mobile phones or networks	2010

Table 2.1. *Progress of biological methods of analysis*

2.2.2. *Foucault's apparatus*

Giorgio Agamben hypothesizes and wants to highlight that the notion of apparatus is a key concept in Foucault's work. Agamben points out that Foucault never provided a precise definition, which would only become available in the 1970s, when the author became interested in governmentality and man's government. The best definition, quoted by Agamben, dates back to the series of seminars held by Foucault in 1977:

> "What I'm trying to pick out with this term is [...] a thoroughly heterogeneous ensemble consisting of discourses, institutions, architectural forms, regulatory decisions, laws, administrative measures, scientific statements, philosophical, moral and philanthropic propositions – in short, the said as much as the unsaid. Such are the elements of the apparatus. The apparatus itself is the system of relations that can be established between these elements" [FOU 94].

Further on, Foucault formulated the following definition:

> "I understand by the term 'apparatus' a sort of – shall we say – formation which has as its most major function at a given historical moment that of responding to an urgent need. The apparatus thus has a dominant strategic function" [FOU 94].

This notion is essential. In the moment of urgent need, it is necessary to educate or "evangelize" users. However, we are not taught in school how to use these apparatuses, it is a learning that takes place between mates or friends, or in the parental relation established between children, in possession of knowledge, and parents. This is also why mentoring is the basis of medical formation.

Two points are significant and worth pointing out. The first is the paramount role played by social relations in the formation of usage, which no one can call into question. The advent of famous social networks (Facebook and Twitter) together with the increasing significance of the mobile phones that support them is now a matter of fact. The diffusion of medical knowledge aside from the doctor's opinion has become a reality. The second important social fact is the inversion of the doctor–patient relationship. The doctor, formerly in possession of knowledge, is here replaced by a patient that gathers more and more knowledge about his own health. This inversion is paralleled by the dismissal of the role of educator normally taken on by healthcare professionals. We cannot be certain whether one phenomenon causes or is caused by the other. Quite succinctly, this author attempts to define the apparatus in three points:

> "It is a heterogeneous set that includes virtually anything, linguistic and nonlinguistic, under the same heading: discourses, institutions, buildings, laws, police measures, philosophical propositions, and so on. The apparatus itself is the network that is established between these elements. The apparatus always has a concrete strategic function and is always located in a power relation. As such, it appears at the intersection of power relations and relations of knowledge" [AGA 07].

Consequently, the analysis of this industrialization will pivot around three major axes. The first one corresponds to institutions and buildings, and then pertains to healthcare sites. The second has to do with myths, fears, the discourse on telemedicine, e-health and then industrialization. Finally, the last one concerns laws and regulations that will favor or try to hinder this inevitable evolution.

This notion of apparatus is at the core of this work. Agamben regards the apparatus as a network between technical equipment and users, a view quite different from the Marxist perspective according to which an object represents crystallized labor. It also introduces the notion that the object considered becomes part of power relations.

2.2.3. *Hyppolite's apparatus*

In his work, Agamben does not hesitate to create a link with Jean Hyppolite's thought. According to Hyppolite, destiny and positivity are two key concepts of Hegel's philosophy. The term "positivity" falls right between natural and positive religion. The latter includes the set of beliefs, rules and rites that are imposed on an individual from external forces in a given society [AGA 07, p. 12]. Can we then claim that there is a positive religion of the medical robot? In any case, the existence of a robot is highlighted by all the establishments that own one, even if they use them to a limited extent.

Hyppolite shows that the difference between nature and positivity corresponds to the dialectics between freedom and constraint. Positivity is a historical element that turns into an obstacle to men's freedom. This distinction is relevant in order to understand the development of medical technologies. Wouldn't constraint be quite evidently then the impossibility for health-related costs to increase?

2.2.4. *The notion of medical apparatus*

A medical apparatus is an appliance, device, implant, reagent for *in vitro* use or a similar and related product employed to diagnose, prevent, or treat diseases or other health conditions. It will not achieve its primary intended action by acting chemically in or on the body, which would turn it into a medication. Medical apparatuses act differently, employing for example medical examination, mechanical or thermic tools (Definition of the Global Harmonization Task Force on Medical Devices (GHTF), part of the International Medical Device Regulators Forum (IMDR)).

Some of these are:

– medical devices as defined by European and American legislation;

– analysis systems, including medical imaging and the materials necessary for biological analysis;

– systems able to perform medical actions, including medical robotics;

– *in vitro* and *in silico* technologies, such as companion diagnostic tests and Point of Care testing;

– prostheses, implants and orthotics, as well as brain-computer interfaces (BCI).

In addition to these systems, we should pay attention to:

– the creation of software necessary to use them, without delving into other kinds of applications, but including communication modes and signal processing methods;

– the optimization of their performances as well as related mathematical algorithms, simulation tools, and the measurement and assessment of elements of the human body.

2.3. The meaning of apparatus

Agamben reminds us that the term "apparatus", as it is commonly understood, can be broken down into three meanings: legal, technological and military [AGA 07, p.19].

2.3.1. *Legal meaning of apparatus*

The apparatus is that part of the judgment that includes the decision as distinct from the reasons. It is the part of the sentence or law that decides and asserts. Thus, medical protocols are parts that are totally integrated into the apparatuses that will be implemented in the field of health. As a result, evidence-based medicine lays the foundations for the aspects necessary to get the industrialization process started.

2.3.2. *Technological meaning of apparatus*

The apparatus is the disposition of the pieces of a machine or mechanism and, in a broader sense, the mechanism itself.

2.3.3. *Military meaning of apparatus*

The apparatus is the set of means available in relation to a plan. The military meaning introduces the notions of tactical strategy and operational level. It is interesting to note that these notions are nearly completely unknown in the medical field. Strategic notions have to do with the banal observation that the rules and decisions concerning the research and evolution of products, markets and technologies vary from one organization to another. Moreover, the notion of strategy refers to the activity itself of an organization. The traditional viewpoint of the high ranks consists in making a distinction between strategy, tactics and the operational level. These concepts were introduced in the military sphere.

There are then three different levels. The policy consists in describing the objectives and goals to be reached. It defines long-term options concerning the battlegrounds of the undertaking, which are generally called "Strategic Business Units" (SBU). In the healthcare field, this definition did not become an industrial reality, which explains why hospitals work with organizations based on the distinction between specialty and disease.

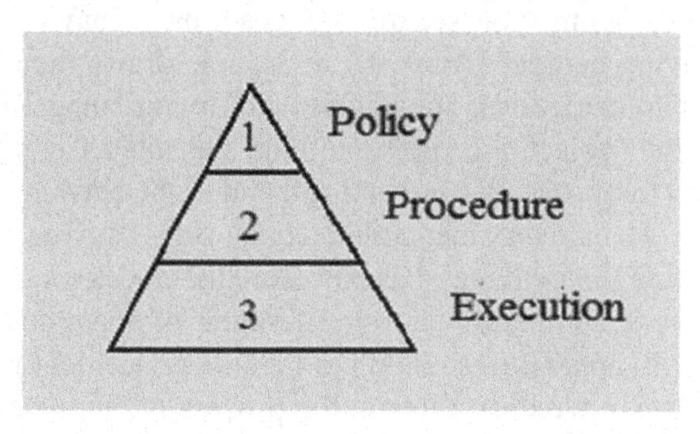

Figure 2.1. *The three levels of action*

The procedure consists in the actual means used to reach these objectives. This is why many authors mention an "organizational level" rather than a "tactical level". Medical protocols are part of this level.

The execution consists in carrying out operations while being aware of the means and of the objectives or goals that have to be reached. Setting up hospital information systems should favor a coherent implementation on this level.

2.3.4. *Economic meaning of apparatus*

Agamben wishes to include a fourth meaning of the term, the so-called "economic" one. He reminds us that the word "oikonomia" means household (Oikos) administration, or more generally management [AGA 07, p. 20]. According to him, "what is common to all these terms is that they refer back to this *oikonomia*, that is, to a set of practices, bodies of knowledge, measures, and institutions that aim to manage, govern, control, orient – in a way that purports to be useful – the behaviors, gestures, and thoughts of human beings" [AGA 07, p. 28]. Thus different options appear. First of all, let us focus on the concept of "biopower", i.e. on how the power of the apparatus acts upon humans and other members of the "bio" world. The other notion derives from the definition of the new offers in terms of treatment sites. The current vocabulary concerning this topic is willingly ambiguous, as is proved by the term "home care".

Agamben comes to the following definition: "I shall call an apparatus anything that has in some way the capacity to capture, orient, determine, intercept, model, control or secure the gestures, behaviors, opinions or discourses of living beings" [AGA 07, p. 31]. He draws the conclusion that there are only two categories: apparatuses and living beings. Between these two, he places the subject. According to him, the subject is the result of the relation and, so to speak, of the hand-to-hand fight between living beings and apparatuses. He adds an incomplete conclusion: "It would probably not be wrong to define the extreme phase of capitalist development in which we live as a massive accumulation and proliferation of apparatuses" [AGA 07, p. 33]. As a result, apparatuses could not be a mere accident men are caught in by chance. They have deep roots in the process of "humanization" and in the development of capitalism. These apparatuses carry out what men, on the other hand, would not be able to do in such circumstances. This is why we need to leave behind the concept of a tool as defined by anthropologists and why the notion of apparatus starts making sense. Thus, robots perform surgery better than men, since the surgeon's shaky hand holding a tool does not come into play.

Agamben mentions the split introduced by the "oikonomia", namely the division between being and action. This scission separates the living being from himself as well as from his relationships with his setting and environment. According to Agamben, this represents "what Uexküll and then Heidegger name the circle of receptors-disinhibitors. The break or interruption of this relationship produces boredom in living beings – that is, the capacity to suspend this immediate relationship with their disinhibitors" [AGA 07, p. 36]. We will have to investigate the importance of this division.

2.3.5. *The religious apparatus*

The religious apparatus introduces the notion of sacrifice, which is what we must do as far as religion is concerned. In the health industry, we have to stretch this notion to ethics. The question becomes interesting. We should notice how medicine has been, for several years and in numerous countries, in the grip of religion. The secularization of medicine began with the French Revolution, although it started being effective only around 1850.

Lucien Lévy-Bruhl regards myths as having a religious origin. According to him, they have to do with a primitive mentality often linked to religious beliefs, dogmas, or at the very least to mythologies on the afterlife and on the origin of beings and objects. Should we consider users as primitive beings or should we regard this philosopher's theory as carrying more weight than it should [LÉV 27, LÉV 22]?

Agamben revisits religion, which he sees as a separation. The apparatus that performs and regulates this separation is sacrifice. There is a shift then from the profane to the sacred, from men to gods, through a series of rituals that the author qualifies as "minute". According to him, these rituals vary in relation to how dissimilar cultures are. Hence, his deduction that capitalism seems to push to the extreme this process of separation, which takes place by means of highly technical apparatuses.

This notion of religiousness of technical processes is evoked by several authors. Angèle Kremer Marietti mentions it: "Communicators are also communicants. They labor together towards the same shores and conquests, in order to engage in tedious and gratifying boardings" [KRE 08, p. 34]. According to this author, the matter amounts to determining these exchange sites and interstices where this communion takes place thanks to technical tools. Communication would then come naturally and, as Dominique Wolton proposes (2005), there would be no need to save communication. We have to notice that the singular conversation of a doctor, the exchange taking place at church during confession, and verbal communication via mobile phones are all part of the same mechanism. This distinction leads us back to our former discussion about the user circles, which represent this progress. Jacques Perriault points out the existence of two kinds of discourses: secular and religious. "One is precise, technical, secular – so to speak – in relation to the abilities of machines. The other is sweeping, ideological, hypnotic, similar to the object ready for religious discourse" [PER 08, pp. 72–3]. Medical imaging significantly figures in both of these discourses. It is in turn magical and only available to specialists. It will enable us, by means of a visual representation and an automated diagnosis, to detect all kinds of female-sex breast cancers before their full development. Only neurology specialists competently trained will be allowed to carry out the diagnosis for a cerebrovascular accident.

2.3.6. *The disciplinary apparatus*

While referring to Foucault's work, Agamben is disquieted by how capitalism generates disciplinary societies by means of apparatuses that aim to create, through a series of practices and discourses, docile, yet free, bodies [AGA 07, p. 42]. Agamben sees then a very wide gap between mobile phones and the screen. The user obtains a number through which he can be controlled, whereas the screen remains passive. An individual that spends his evenings in front of a television will only gain the frustrating title of couch potato and his inclusion in an index (viewing figures). Agamben points out that "here lies the vanity of the well-meaning discourse on technology: [it] asserts that the problem with apparatuses can be reduced to the question of their correct use". He reminds those promoting such discourses on the consequences of the media apparatus that they are actually captured in it [AGA 07, p. 45]. This notion of non-use corresponds to the one advanced by Perriault when he claims that "the term 'use' initially defined social practices supposed to be good form, as it were". This social-practice-based approach is clearly a development of the anthropological one devised by Bourdieu [BOU 82, BOU 03], Hymes [HYM 81, HYM 86] and Leroi-Gourhand [LER 64, LER 65].

Agamben concludes his work by analyzing the concept of power interactions: "The more apparatuses pervade and disseminate their power in every field of life, the more a government will find itself faced with an elusive element, which seems to escape its grasp the more it docilely submits to it" [AGA 07, p. 49].

In this work, we will then leave behind Pierre Schaeffer's approach, namely the "Machine for Communicating", and base our analysis on the notion of apparatus.

2.4. The "plus" man

The "plus" man is born thinking that we will add or integrate into men elements that did not previously exist. Thanks to these elements, men will be more likely to remain in good health.

2.4.1. *The plus man, a product of cybernetics*

Enhanced man, android, bionic, cyborg, prostheses, gerontechnology, robolution, anthropotechnics and transhumanism are all terms that pertain to the cybernetic revolution and to the theory of technological singularity.

This theory was described by such writers as Irving John Good and Vernor Vinge and advanced by Ray Kurzweil.

DEFINITION 2.2.– *Singularity is the moment in future when the advanced technologies built by mankind are so operational that men will no longer be able to understand them, and when these technologies exceed humans or steal their role in the hierarchy of species.*

According to Kurzweil, our world will remain human but will also move past our mere biological origins. It is the aforementioned apparatuses that will outdo mankind. Behind these ideas, we can find then the cybernetic revolution.

2.4.2. *The production of quality of life*

Quality of life, as an event, only happened recently. It is much harder to bring about quality of life than to treat, since we add the complexity of man to the intricacies of medicine. The production of quality of life is also trickier to assess in terms of clinical trial and evidence-based medicine. Other social, financial, technological and political factors can have more of an impact on quality of life than medicine, which is thus relegated to actions of second order. Finally, medicine has to frequently manage situations where an increase in the quantity of life necessarily entails a drop in quality of life.

2.4.3. *Incorporation: tools turning the tables on men*

Following this line of thought based on the notion of "tool" consists in wondering how medicine will develop systems which become artificial extensions of humans. We recall then De Kerchhove's exteriority theory. These prostheses are becoming more and more real. "People are beginning to treat their mobiles as human beings because they symbolize contact, friendship and attention" commented Dr David Nott, addiction specialist at the Priory Hospital in Southampton (southern England[4]). It is very likely that the elements created by this industrialization will be seen as integral parts of men, with the creation of a new prosthetic man. If we follow this approach, medicine becomes a prosthesis factory. Some are communicational, such as mobile phones, others are physical, such as the exoskeletons that allow people to walk or to learn to walk again, and others again are physiological, such as insulin pumps.

2.4.4. *The prosthetic or orthotic man*

The objects that belong to the industrial world are prostheses that allow men to be more efficient. Men fall within a "man-technology-object-environment" continuum that emphasizes the "technology-object" link. In this case, technology and boredom are related and they even seem to gain strength from each other, which would explain the constant need for innovation and better technology to increase efficiency even more. If this boredom theory is applied to our mobile phone, we are led to believe that the object itself participates to the performative character of boredom. The modern man needs this tool, which actually becomes a "necessary prosthesis", because he becomes bored. However, the tool turns into an orthosis rather than a prosthesis. An orthosis is a device that compensates for an absent or deficient function. The absent or deficient function would then be the managing of boredom. In order to deal with it, it is necessary to have a time base whose former peripheral function, the clock, is the necessary link.

Man is a being fitted with several prostheses. Let us consider some examples. Cars are natural extensions of our body's limitations in terms of natural movement. They allow us to go farther and faster. We would struggle to imagine a world without cars. The car example is interesting since we can use it to show the strange links created between technology, prosthesis and death. Every kind of technology becomes dangerous for men. It necessitates a search for safety measures and confidence-building. It lays bare the fundamental mortality of men, which can however be treated statistically. This is certainly why we try to conceal the most "human" prostheses, such as hearing aids or articulable limbs. It is possible to discern the myth about the two possible kinds of discharge from hospital in this relationship between prosthesis and death. The former consists in either actual death or the

4 this is a report on the addiction to mobile phones broadcast by AFP on Thursday 26 January 2006.

"partial death" of being discharged with a prosthesis. The latter is the healthy discharge we mentioned at the beginning of the book. This situation makes prostheses one of the key elements for the treatment of chronic diseases. Hence, the research on artificial hearts and pancreases or on cochlear implants is conducted.

Lars Fr. H. Svendsen claims that "anthropocentrism gave rise to boredom, and when anthropomorphism was replaced by technocentrism, boredom became even more profound" [SVE 03, p. 123]. According to him, this happens because technology introduces a dematerialization of the world in which things disappear in their mere functionality. The information society has intensified this phenomenon and devices become outdated even before potential users can learn to make use of them.

2.5. Science, technology, art and industrialization

Jean Marc Lévy Leblond estimates that science has been around for only a few centuries, whereas technology has been a thing for a long time. We can define science as a form of rational and experimental knowledge. Elaborate forms of transmission are necessary for the subsistence of science. It is also necessary for this knowledge to be "inseparable from its immediate finalities, of symbolic, religious or divinatory order". Science appears as soon as "theoretical knowledge is detached from its empirical context and from its sources and applications" [LEV 13, p. 13]. The famous formulation of Descartes's Discourse on the Method states that, thanks to science, we can "thereby make ourselves, as it were, the lords and masters of nature". Descartes explains how, thanks to science, men will benefit and be able to remain in good health. Technology is possible by means of human experimentation. However, the link between technology and science is not direct.

The man who built the first steam engines, Thomas Newcomen, knew nothing about the scientific theory published in Latin by Denis Papin. Unlike technology, science is born within culture and it is one of its components.

2.5.1. *A current kind of medicine stemming from technology*

We do not mean to belittle the actors of current medicine, but their practices are closer to technology and art than they are to science. Art shares more with technology than it does with science. The result is that industrialization pertains more to art than science. Jacques Ellul explained how technology would turn into industrialization, which was a process where, autonomously and inexorably, the means replaced the ends. On the other hand, medicine had been using culture as its source for its technical terminology derived from Latin and Ancient Greek for several years before these subjects were dropped in schools. The inexorable consequence was that this kind of terminology was used less and less when naming diseases and practices. The result is an e-scientificity that characterizes the new names given to medical practices and concepts, which is linked to a necessary process of translation that changes AIDS into SIDA!

Romanticism is the period in which, for the first time, we observe a virulent reaction against the dominion that technology and industry start to exert over the world. We will focus on this point in the next chapter.

2.5.2. *Science and technology in the face of the industrialization of health*

As Lévy Leblond's history-related examples show, it is possible to possess technology without science and vice versa. As of now, the industrialization of medicine, and consequently the expression of the current aims of the medical technique through means and the appearance of knowledge in the same field, are taking place simultaneously. The relationship between genes and viruses in the development of certain kinds of cancers and the progress made in genetics and its associated manipulations are two examples that illustrate this situation.

As a result, this industrialization of medicine can be brought about by:

– technology: robotization, the appearance of biological diagnostic apparatuses, and more and more effective medical imaging;

– science: thanks to genetics, proteomics, and any science ending in "omics";

– the growth of the health field;

– technology and science combined.

Sci-fi authors have imagined technologies that are as promising as they are extraordinary since the dawn of time. The analysis conducted by scientists on these works turns out to be an excellent means of innovation. Don't we often say that Jules Verne's ideas have become a reality? Science feeds on imagination and speculations as well. Thus, we try to obtain materials whose properties correspond to health needs, for example to build new prostheses or systems able to pick up muscular energy and recycle it to other parts of the body.

3

Industrialization: its Obstacles and its Rules

The aspect most often associated with industrialization is undoubtedly the opposition to change. That is why it is necessary to reflect on the notion of opposition as well as on the impact of the characteristics that we expect to define this process.

3.1. The opposition of the actors in the health industry as an obstacle

This opposition to change is amplified, on the one hand, by a bi-individual vision of treatment that needs to be agreed upon more and more and, on the other hand, a collective vision of healthcare dominated by such goals as public health and optimization.

3.1.1. *The double level of opposition*

The individual vision is characterized by precise actions performed by the healthcare professional, namely the provider, who struggles to treat the patient or to rehabilitate the disabled individual, whom we will call the sick person. Dialogue is necessary in this relationship. The collective vision implements accreditations, certifications, norms, procedures and protocols that do not visibly affect the development of the disease or the rehabilitation of the disability. The joint interest of the provider and of the sick person consists in increasing the effectiveness of treatment in order to obtain the result expected. The collective interest defined in the word "public health" consists of security and control. It is therefore a matter of conflicting objectives.

3.1.2. *The notion of opposition*

The notion of opposition is significant in the context of the health industry. Liette Lapointe analyzes different forms of opposition. We have chosen not to discuss them in detail. However, we should point out, at this stage opposition emerges mainly because the distribution of power is upset and not enough attention is paid to the specific needs and interests of the groups of actors involved. The only solution for the development of medical industrialization consists in reducing the risk of perturbations linked to new power. According to some, this entails the formation of communities, since they reintroduce an element of specific interest for the individuals, which consequently makes it a successful policy.

3.1.3. *The role of "implementers"*

The problems associated with users' opposition are real and complicate the task of organizations. However, the role played by "implementers" – whether they are IT experts, advisors as well as producers, software developers, hospital builders or administrators such as operators – has been only studied superficially or ignored altogether.

The rare studies that deal with the implementers' actions/reactions mainly emphasize how opposition can be prevented, which is not what concerns our current actors.

Two questions become paramount.

Question 1: How do "implementers" react when faced with the users' opposition?

A taxonomy of replies can be discovered in the series of users' behaviors, which we will do by analyzing the different uses, voices, data, images and positioning stemming from Abraham Moles' approach. Liette Lapointe has listed the majority of these uses.

Question 2: What are the effects of the implementers' reactions on the users' adoption or opposition?

We will therefore have to study the influence of these reactions on the intensity of the users' behaviors. These reactions change according to the level we consider: individual, group or organization.

Apathy	Inaction [MEI 95]
	Distance [GAB 99]
	Lack of interest [COE 99]
Passive opposition	Creating waiting times [TEP 94]
	Refusing responsibilities [AGO 97]
	Humor [JUD 81]
	Leaving or being transferred [CAR 86]
Active opposition	Reporting [TEP 94]
	Rumors [AGO 97]
	Coalitions [CAR 86]
	"Peaceful" demonstrations [JUD 81]
Aggressive opposition	Making mistakes [JUD 81]
	Violence, rebellion [WIC 98]
	Sabotage [GLA 99]
	Destruction, murders, terrorism [COE 99]

Table 3.1. *Table of the kinds of opposition according to Liette Lapointe*

Liette Lapointe summed up the causes of opposition, as they are presented in the table below.

Individual	Fear of the unknown [ZAL 77]
	Little tolerance [KOT 79]
	Habits [HEL 83]
	Feelings of powerlessness [ABD 95]
	Ignorance [AYR 78]
Group	Cultural values [COE 93]
	"Class consciousness" [MAR 67]
	Group dynamics [COE 93]
	Expectations of a given group [JOS 91]
Organization	Remuneration systems [KOT 95]
	Setting up [KRE 92]
	Forms of communication [KOT 79, GRI 93]
	Structure and culture [MIN 79]
	Distribution of power [MAR 83]
	Sharing of responsibilities [KIN 89]

Table 3.2. *Forms and levels of opposition according to Liette Lapointe*

For each of the abovementioned elements, it should be possible to analyze how these factors have played a role without contributing to opposition. However, we will not delve into this.

The same can be said for her table concerning perceived threats. Further on we will come back to the notion of loss of power and to Foucault's approaches in order to discuss new forms of power.

Economic loss	Hellriegel, Slocum and Woodman [HEL 83]; Zaltman and Duncan [ZAL 77]
Psychological distress	Kreitner [KRE 92]; Diamond [DIA 89]; Diamond [DIA 93]; Marakas and Hornik [MAR 96]
Loss of control	Hellriegel, Slocum and Woodman [HEL 83]; Tichy [TIC 83], Kanter [KAN 95]
Restructuration of work	Jermier, Knights and Nord [JER 94]
Splitting up of work teams	Kreitner [KRE 92]; Griffin [GRI 93]; Aldag and Stearns [ALD 91]; Schermerhorn [SCH 89]
Injustices	Greenberg [GRE 90]; Joshi [JOS 91]
Poor results	Kotter and Schlesinger [KOT 79]; Willer [WIL 81]
Fear of failure	Kreitner [KRE 92]; Dubrin and Ireland [DUB 93]
Decreased status	Kreitner [KRE 92]; Schermerhorn [SCH 89]; Dubrin and Ireland [DUB 93]
Loss of power	Foucault [FOU 80]; Markus [MAR 83]; Tichy [TIC 83]

Table 3.3. *Table of perceived threats according to Liette Lapointe*

3.2. A comparison with other economic sectors

The comparison with other economic sectors points out quite aptly how necessary the industrialization of medicine is.

3.2.1. *Managing the complexity of health*

Medicine is a complex system acting on another complex system which is, as a matter of fact, man. It would be possible to confine our analysis to man's health, but we would simply reduce two complex systems into one. A complex system is made up of a very large number of interconnections and interactivities. Its internal organization is often dependent on the forces existing between parts.

A complex system is defined as a set of several parts, each of them interacting with the others and with the environment.

The industrial domain is characterized by its management of complexity. For example, any kind of factory owner has to manage and follow a network of suppliers in relation to the different kind of demand from the clients. In the health industry, it was noticed that a GP manages, on average, 180 different activities per patient a day. In the American Medicare system, a patient deals with 229 doctors in his lifetime. "Therefore a doctor can prescribe one of the 7,200 medical acts liable to itemized deduction, one of 5,000 medications, one of the 50,000 medical apparatuses available (materials, implants, etc.), he can send a patient to a specialist, prescribe biological tests, x-rays, he can have a patient hospitalized, he can ask to be assisted by paramedics, nurses, physiotherapists, dentists, etc.". Such is the situation in France as described by Laurent Degos [DEG 13, p. 15]. We also have to point out that the number of combinations is not comparable to the rational products of this industry. This situation justifies medical expertise. We should however wonder about the consequences of industrialization on the number of possible combinations.

3.2.2. *Adjusting to patients (the clients)*

Industry aims to provide a client with some goods or a service. In the industrial domain as much as in the service one, everything is normally done in order to adapt to the clients' needs and expectations, and, in some cases, to respect their specific demand. However, in the health industry the patient has replaced the client for several years. Thus, he receives a prescription, which is an order he is demanded to follow. Currently, some actors in the health industry are becoming interested in the notion of observance, since patients have decided not to respect these orders any longer.

In the United States, less than half of all patients receive clear information on the effects and consequences of treatments. In Europe, laws on medical information such as the Kouchner Law in France should prevent such figures but surveys show that this is not the case. Recent scandals, like the one created by fourth generation COCPs, can only substantiate that. In our Western countries, surveys show that less than half of all patients are satisfied with the levels of control they exert on medical decisions.

As we formerly mentioned in the introduction, life expectancy is the most useful indicator. However, it does not take diseases into account. This is why we have seen the appearance of notions of declared morbidity. In order to analyze these situations, establishments have asked patients to state their pathologies. The World Health Organization (WHO) makes use of these statements to gauge the health condition of a country. An industrialized kind of health could use them to promote technological solutions. This international organization has developed HALE (Health-Adjusted Life Expectancy), a measurement of life expectancy in relation to health conditions including such parameters as mobility, cognition or depressive states.

Another indicator adjusted to the patient is life expectancy after diseases have been found. It is the reply to a question frequently asked by the sick person. It is also the first, and perhaps the only, mainstay of the role of *santacteur*.

3.2.3. *Increasing effectiveness and reducing losses*

Industry "monitors" its processes in order to enhance quality, identify ineffectiveness and reduce losses and bad production. In the health industry, we estimate that more than a third of health-related expenses do not actually contribute to health. A good example is the ambulance transport service.

This increased effectiveness requires us to agree on the evidence. Several pathologies are still being treated according to the opinion of specialists and experts, which constitutes the first level of evidence. The second level has to do with conferences held to reach consensus, experts' meetings or nearly perfect searches through the literature. The third level consists of clinical trials, which include randomized, double-blind and placebo-controlled trials. Statisticians call this experimental design. Lastly, the fourth level consists of a meta-analysis of all the previous studies. These approaches have entailed certain forms of excess, such as treating healthy individuals who presented risk factors, which might have resulted in a reduction of their life expectancy.

3.2.4. *Creating a team for auditing and controlling healthcare production*

In the industrial field, teams focused on auditing and controlling are constantly at work. The workers, technicians and engineers are coordinated so that effectiveness can be attained. They are also helped both when operations are being started and during their task. In the health industry, in Europe as much as in the United States, more than 50% of adult patients report problems in terms of coordination of treatment, results and diagnoses that are not notified, and difficulties in communicating with doctors. Around 30% of these patients report the presence of an ineffective healthcare organization.

This approach based on control consists of carrying out case-control studies or cohort studies. Large samples of population are compared while taking into account the presence or absence of a risk factor which therefore defines different classes.

These cohort studies are not unbiased and provoke controversies among experts. It is a matter of knowing how to measure survival. If we take the example of a tumor, we have to consider:

– survival without a noticeable progression of the tumor;

– survival with the regression of the tumor and thus a clinical reaction; and

– survival with a biological response characterized by a drop in the diagnostic marker for cancer.

We should add to these three types of survival some life-related criteria, such as:

– living a life with or without a psychologically or physically disabling ablation;

– living a life that requires or does not necessitate prostheses in order for our health condition to improve.

3.2.5. *Reducing the number of medical mistakes*

The reduction in the number of mistakes and nosocomial diseases has become a real taboo subject. On average, hospitals in Europe kill more than car accidents. In the transport sector, manufacturing industries as well as airlines have learned from the past. Procedures, particularly those of preventive maintenance, are adjusted to ensure excellent aviation safety. This is not the case for the health industry, where 20% of patients in the Medicare system are re-hospitalized 30 days after their discharge and a third of patients have somehow suffered during their stay. According to a poll called ENEIS, 7% of hospitalizations involve an unwelcome incident and 1% of them are fatal. This would affect, in France, 150,000 people a year, and therefore involve a few more than 10,000 cases that end with the patient's death and around 2,000 accelerated deaths. These figures are three times higher than those related to motor vehicle deaths. This is why we find the expression "hospital violence" quite apt, by analogy with the phrase "road violence".

Biochemical diagnoses have simultaneously become the first step towards the industrialization of medicine and entailed a remarkable alteration of the notion of disease. There is a difference between being a sick person because we present the evident symptoms of a disease which disrupt our daily life and being sick because we have a chemical anomaly or an abnormal biological constant. This leads us straight back to the problem of normality (what is normal?). If we hypothesize that each individual is different, then the status of "disease" becomes chronological because we have established the concept of future and uncertain disease.

The second approach consists in reducing involuntary mistakes and undesirable incidents. These happen quite frequently and take place even with the best procedures and highest levels of expertise. These mistakes are more frequent than car accidents. They would be the equivalent of a plane crash per day, if we compared these figures to those concerning civil aviation. Moreover, any kind of technology involves its share of accidents. In medicine, the progress done in terms of transplants is undeniable but it was still necessary to find a solution, which is not always perfect, against transplant rejection.

3.2.6. *The use of IT*

When at the bank, all customers can see the history of their financial operations and also check their bank balance online. They can even make bank transactions online. According to the survey conducted by the National Academy of Medicine in the United States, 20% of biological analysis results and medical data are not sent by the laboratory to the healthcare center on time. Besides, 25% of healthcare professionals have to order tests to improve diagnoses.

3.2.7. *The transparency of medical services and their prices*

In France, when you buy consumer goods or book services, you can compare the prices and find out about the effectiveness of these products and services. When it comes to the French health industry, none of this holds true.

In the United States, 63% of patients are not aware of the cost of their healthcare and do not know the amount they will have to pay before they receive the invoice; 10% of patients will never discover the cost of their treatment.

In France, under the organization of social security, transparency is non-existent as soon as the regime of the third-party payer is implemented. The question of the co-payment becomes then the real deal-breaker. It supposedly reduces the costs of healthcare expenses, even if no one has any idea about the total amount. The knowledge of the treatments employed is also quite unclear, whether because of the complex medical jargon used or the omnipresence of the disease for each individual.

3.2.8. *The complex adaptable system of health*

Health is currently becoming more than a mere complex system; it is becoming a complex adaptable system that also endeavors to optimize communal values. It may be a matter of costs reduction, performances and adjustment to every individual. However, we should not forget about the role played by the enrichment of healthcare professionals, their reputation, or the growth of a technology or medical approach. This adaptable system evolves into two natural adjustments, summarized too concisely by Charles Darwin's theory of evolution of the living world and Gregor Mendel's theory of gene inheritance.

4

Acceptability and Diffusion

Usage is a key element of the process of industrialization. However, this process requires more. Aristotle teaches us how all human beings aspire to knowledge, but he could not discuss industrialization, which is a concept that did not exist in his time. We have seen that diagnostic methods increase our knowledge about diseases. A technical solution of any kind can only be industrialized if it becomes useful.

4.1. The criteria of the analysis

In the case of a complex system, we are tempted to use a technical solution. There are many theoretical ways of getting to this point, but it is easier to understand how uses are established and to start from there.

4.1.1. *The interventionist approach of health policies*

The first temptation is to require the health actors to act. However, the health industry is far from a disciplinary Foucauldian system. As a consequence, this solution is not very effective. Nonetheless, the high ranks responsible for health in every country attempt to do this and we know the results. The dream of knowledgeable medical societies is to impose protocols as well as to ensure their validity.

The second temptation has to do with external management, based on promptings or common objectives. In most cases, the result is not as expected and sometimes it is even contrary to the objective. In England, the temptation to implement performance-based forms of payment for doctors had the opposite effect in terms of people treated or dropped in mortality. Management has not bettered the indicators of public health.

Finally, the last solution conceived consists of observing our neighbor and copying him. Once again, this generally does not work, since it is quite unlikely that the complexity of the Swedish healthcare system corresponds to that of the French one. Sometimes, even the fundamental values themselves are involved. In Great Britain, it is acceptable to opt for a utilitarian method in order to assess the use of a medication or therapy on a patient. In France, the most important value is quality in terms of access to all forms of treatment.

Complex systems have a natural tendency to self-organization. However, we are trying to pilot the health industry and, as a consequence, the choice of core values and of their translation into goals becomes essential.

4.1.2. *The usage-based approach in the health field*

We generally use five criteria to analyze the suitability of a technological tool: acceptability, accessibility, learnability, usability and usefulness.

Term	Content
Acceptability	Does that cater to one of my daily needs?
Accessibility	Can I actually make use of it?
Learnability	Can I try it and learn how to use it?
Usability	In which circumstances can I make use of it?
Usefulness	What can I use it for?

Table 4.1. *Different kinds of criteria for the suitability of a technological tool*

One of the practical difficulties consists of the fact that we have to deal with two complex systems that can respond to the issues presented by the individual and medicine. As a result, we obtain at most incomprehension and often clashes. What is more, if I only take into account medicine, nothing tells me that the doctor and the nurse will share the same opinion on this issue.

4.1.3. *From usage to knowledge*

DEFINITION 4.1.– *Usage is how we employ something, which corresponds to a service or to an application.*

The types of communication correspond to the set of technical means that allow the diffusion of messages. It is a more general process. Men, in order to communicate, can use machines in which “memory” is just one of many functions. Usage in the medical field has been defined in relation to treatment protocols more or less observed. We can safely bet that industrialization will present them with a new challenge.

DEFINITION 4.2.– *Memory is a physical device that allows the conservation and restitution of information or data.*

Aristotle regarded memory as part of the chain that leads to the acquisition of knowledge. Knowledge, whether medical or not, is rooted in perceptions. They play such a significant role that certain doctors refuse telemedicine on the sole basis that they will not be able to listen to the patient’s chest and produce a good diagnosis. These perceptions, often linked to incidents, lead to the constitution of memories. These acts of memorization give birth to experience. Experience is a form of factual knowledge, but it will not be able to answer the question “why”, which is necessary for industrialization to take place.

4.1.4. *The special features of health-related IT*

One of the characteristics of ICT is the interaction it enables. Let us point out that this interaction often takes place in healthcare procedures, which would *de facto* justify their use.

DEFINITION 4.3.– *Interaction is the term used to designate simultaneously the kind of dialog that takes place between user and device and the fact that several people can communicate by means of that device.*

This notion of interaction refers to a discussion on the passivity of the user in the process of communication. These technologies are only used if someone wants to make use of them and if the result is a sort of individualization. Intention cannot stem from traditional and historical power, so it establishes itself through lateral power. In the health industry, this is represented by the new kind of power exerted by patients. If they were at first informed, they are becoming informers and establishing themselves little by little in the leading bodies of healthcare sites, such as hospitals. They will certainly become powerful with the advent of the third industrial revolution. It is not possible to base our study on only one theory of intention. Let us recall the definition of this term.

DEFINITION 4.4.– *Intention is the reason why we communicate or why an individual acts. It is generally attributed to the recipient.*

We can imagine the singular conversation between a doctor and his patient as a kind of intention. Nowadays, it is evident that this relationship is going to change. However, healthcare procedures can be analyzed as a process of individualization, a concept that we will remember. Intention is the source of the kind of trust that we will deal with again later on.

DEFINITION 4.5.– *Individualization is a process that takes place when individuals use these technologies by means of a particular interaction.*

In order for this individualization to take place, we must make reference to phenomena related to our identities. All of us must possess an identity that allows us to separate our civil status from our personal status. We will try to qualify this process of individualization. In the medical field, the notion of translational medicine, in which healthcare is adjusted as much as possible to the patient, pertains to this process of individualization. It is also the desire to move quickly from laboratory-produced scientific discoveries used in clinical medicine to the "patient's bedside" and vice versa.

4.1.5. *From health knowledge to industrialized science*

The key issue has to do with medical science, which will bring about the process of industrialization. We must accept the idea that what keeps men in good health is mainly experiment, and not science. Therefore, we must give an explanation. The shift from experience to science ("episteme" in Greek) is a difficult phase. It is no longer a matter of recording facts and events; we have to explain them by employing general laws.

DEFINITION 4.6.– *Metaphysics is the science that treats human beings as beings and that tries to deduce all the characteristics and features that belong to every existing entity.*

In the health industry, difficulties often arise because elements specific to either metaphysics or science are jumbled together. This confusion is largely maintained in medical training. The complexity of health is partly a result of this difficulty. Another source of confusion is the disappearance of symptom-like diseases in favor of diseases based on a biochemical diagnosis.

The second difficulty derives from the confusion between epistemology and theology. The latter is concerned with objects that do not exist in nature. This confusion, which dates back to the religious origins of medicine, has supposedly started to disappear ever since the secularization of medicine.

4.1.6. *The sources of medical innovation*

We should not forget that we have been able to make progress by learning from our mistakes. In the health industry, certain elements have been isolated so that they could be made safer. Thus, complex systems have become complicated systems. Anesthesia and radiotherapy have become quite safe. The same can be said about blood transfusions, a field that made most of its mistakes by introducing easily testable blood types.

In order to boost progress, researchers want to move from one complex system to complicated systems, even if this involves isolating them. In a complex system, only experiments enable us to know *a posteriori* the consequences of an action, which could have an effect even a dozen years later. This is the case for the impact of certain medications, the long-term effects of which can only be observed in clinical trials. This is also why legislators wanted to supervise experiments by means of such laws as the Huriet law, then the Jardé law in France, or the Temporary Authorization for Use (TAU) for medication or medical protocol testing. If it is not possible to shift to a complicated system that presents predictable consequences, experiment reduces the risks and gives an impression of safety.

Technological innovations are spreading through our society by following specific modes or models. As for errors, biological and medical sciences work according to a "description–understanding–repair" cycle, which therefore becomes the feature of health-related innovation.

DEFINITION 4.7.– *The diffusion of an innovation is the process by which the members of a society get to know and assess an innovation, and finally adopt it. This process involves means of communication.*

This definition implies that the subsistence of innovation depends on the solution of a problem: the expected diffusion has to be permitted and assessable. In the health industry, this situation becomes impossible since the state of health is not quantifiable and results from a tripartite subjectivity:

– the one of the doctors who diagnoses;

– the one of the patients who feels, or does not feel, sick;

– the one of healthcare policies, which regard this or that condition in a patient as the symptoms of a disease that needs fighting against.

In the process of diffusion of medical innovation, we have to point out how our current society, based on fear or anxiety, generates specific behaviors. This promotion of anxiety requires the scientific reliability represented by universities, the medical profession, health authorities and the state. As this promotion of the disease takes places as a pharmacotherapeutical solution or related technology emerges, a lot of optimism ensues.

The model of adoption most often mentioned in marketing books takes up the economic notion of Schumpeter's [SCH 43] gale of creative destruction. According to this theory, products are sold in relation to a "bell model" whose parameters only need adjusting. Researchers in different fields have addressed this concept of diffusion of communication tools. More recently, new approaches have been defined: the "Diffusion of Innovation Theory", the "Theory of Reasoned Action" and the "Theory of Planned Behavior" are some of the examples. These models have been devised to explain and verify those factors that are responsible for the adoption of technologies. We will briefly recall the basic notions of this field.

4.1.7. *The traditional approach: diffusion of innovation (DOI)*

Rogers [ROG 95] defines diffusion as "the process by which an innovation is communicated through certain channels over time among the members of a social system" [ROG 95, p. 5]. He then describes five criteria: relative advantage, compatibility, complexity, triability and observability.

Rogers introduced the notion of opinion leader, i.e. someone who allows information to be diffused. The five aforementioned criteria also refer to some of the key notions that we will tackle later on. The concept of real use of an object will lead us to the study of misuse. The idea of critical mass of buyers is not too useful for our product owing to the rapidity of diffusion. Finally, another key notion is the degree of user-related reinvention, which allows us to put forward the hypothesis that it could bring about a new era and society.

4.1.8. *Relative advantage*

"Relative advantage" defines the degree of improvement with which an innovation is associated. Rogers suggests then that the greater the relative advantage of an innovation, the quicker its rate of adoption.

This notion of relative advantage has been used widely in the health industry to shift from one medical protocol to the other or to accept a new kind of medication.

4.1.9. *Compatibility*

Compatibility corresponds to the degree of conformity to existing values, previous experiences and potential adopters' needs with which the innovation is associated. Tornatzky and Klein [TOR 82], as well as Ndubinsi and Sinti [NDU 06], made this remark. Any innovation was in fact more easily adopted when it turned out to be compatible with the potential adopter's usage.

4.1.10. *Complexity*

According to Rogers, complexity depends on the degree to which an innovation is perceived as hard to understand and employ. He suggests that an innovation has to be easily understandable for the potential users to adopt it. Lederer and Alii [LED 00] showed that complexity has been able to negatively affect the adoption of the Internet, for example from 1995 up to now, because of "http://" commands, among other things. The reduced complexity, involved in manufacturing, diffusing and using, is one of the main signs of industrialization.

The following two criteria are quite different and cannot be applied to the health industry.

4.1.11. *Triability*

The possibility of experimenting is very significant. Triability is the degree to which an innovation can be experimented on a limited basis. This is the reason why the "free" part is important for TIC-related innovations. Agarwal and Prasad [AGA 98], as well as Tan and Teo[1] [TAN 00], suggest that this triability helps reduce the unknown, and its related fears, as much as possible.

How to try surgical procedures? In most cases they are irreversible.

4.1.12. *Observability*

Observability is defined as the degree to which the impact, effects or results of an innovation are visible. Other authors might prefer the use of the term "demonstrability", following Moore and Benbasat [MOO 91].

Once again, it is difficult to observe the effect of a medication long-term. Clinical studies are generally representative of a given period.

Rogers was the first to introduce the term "early adopter". This approach points out the role played by advertising, which is an element of mix marketing. It must present the product to the consumer by showing how its users make use of it. One of the difficulties linked to that period is that it was then absolutely impossible to define what these uses were.

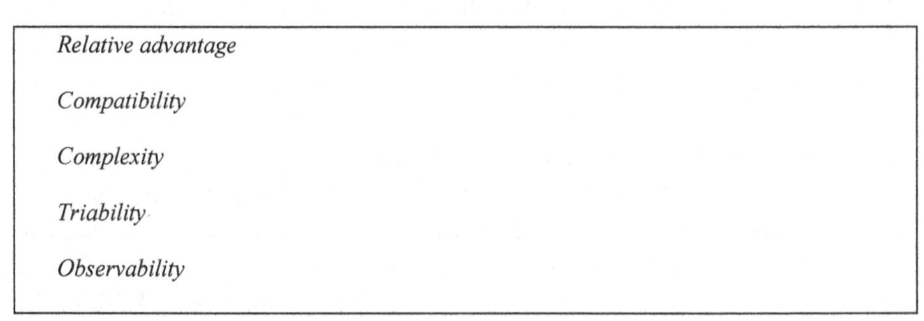

Relative advantage
Compatibility
Complexity
Triability
Observability

Source: Rogers, Bonne question, 1995.

Table 4.2. *Rogers' criteria*

1 These authors insist on the trust we must have to adopt and how the possibility of experimenting the facility.

As for the notion of segmentation, we have to point out at this stage that certain sociologists specialized in usage think that it includes the basic principles of marketing. So, Woolgard [WOO 91] claims that designers assess "the usability" of technical objects before launching them into the market. Thus, the conception of the technical object has to be linked to a "configuration of the user". Besides, these authors misuse the terminology employed, which is not the case with Rogers.

Madeleine Akrich [AKR 87] focuses on the design of technical objects. According to her, this kind of object involves a script that works as a scenario and storyline. This scenario predicts the users' needs by integrating their interests, competences, abilities and ways of behaving. In order to resist this object, Madeleine Akrich comes up with a new idea, in relation to Anne-Marie Laulan, by offering the user to "unsubscribe".

4.2. The models of adoption of medical technologies

Different models of adoption have been proposed for several years and are by now completely adjusted to healthcare products and services.

4.2.1. *The TRA (Theory of Reasoned Action) model*

The TRA model was proposed in 1975 by Fishbein and Azjen [FIS 75][2]. It focuses on the construction of a system of observation of two groups of variables, which are:

– attitudes defined as a positive or negative feeling in relation to the achievement of an objective;

– subjective norms, which are the very representations of the individuals' perception in relation to the ability of reaching those goals with the product.

These authors have emphasized the importance of intention more than the reality of usage. Actually, people who buy something do it in relation to what they feel like doing and not really because of an actual need related to the model they belong to. In the context of IT, this approach does not work.

2 According to the theory of motivated action proposed by Fishbein and Ajzen [FIS 75], beliefs, attitudes and intentions create a causal chain; beliefs lead to attitudes and attitudes in turn lead to intensions and therefore to behaviors.

We consequently notice the inadequacy of this model in relation to the topic of our study.

4.2.2. *The TPB (Theory of Planned Behavior) model*

This model is a development of the TRA one, which is devised to overcome the shortcomings of the previous methodology. Ajzen [AJZ 91] defined this variant as the construction of a form of behavioral control perceived in those situations, where individuals lack substantial ("actual") control on specific behaviors. On the contrary, this implies that there are compulsory behaviors [AJZ 91].

4.2.3. *The TAM (Technology Acceptance Model) model*

Davis [DAV 89] proposed the model called "TAM" in order to explain the users' acceptance of ICT[3]. To that end, Davis proposes a list of variables that define a perceived usefulness and an equally perceived ease-of-use. These two variables are used to define behavioral intentions. Once again, these two elements depend on considerations concerning costs and benefits linked to such systems. Afterwards, studies employing this approach added the notion of "perceived enjoyment" (PE). PE represents the enjoyment of pleasure and the satisfaction associated with a certain behavior [DAV 89].

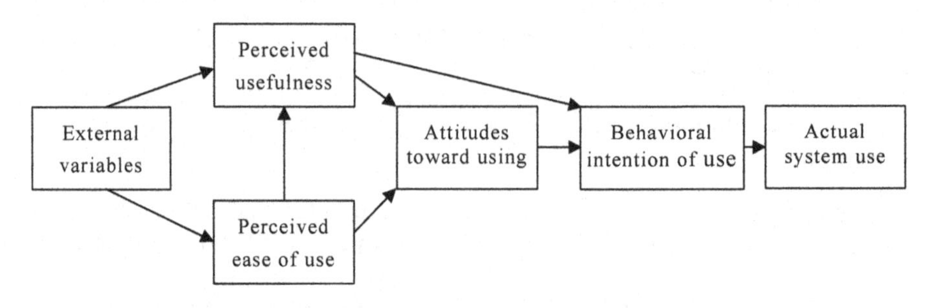

Figure 4.1. *The TAM model (according to [DAV 89])*

3 Davis started with this observation: attitudes towards use and intentions in relation to the use of a technology can be determined incorrectly. Therefore, actual use does not necessarily have to be a direct or immediate consequence of these attitudes and intentions.

4.2.4. *The unifying UTAUT (Unified Theory of Acceptance and Use of Technology) model*

After the criticisms linked to the previous models, Ventakesh [VEN 00] carried out research in order to find a unified model. The UTAUT model also accepts to comply with a motivational model and a "social knowledge" approach. It includes two kinds of variables: those that influence behavioral intentions and those that define actual uses. The author places in the first group: efficiency expectations, effort expectations, social influence and facilitating conditions. In the second group of variables he classes: age, sex, experience and "willingness" of use.

The expectation-based approach turned out to be ineffective. In fact, the future user of a healthcare system is more interested in actually being treated than in defining his expectations of this system.

This model introduces some notions that will be at the core of our discussion. We come back to a form of adoption that requires the sharing of knowledge and, often, competences. By competence we mean "I know how to use it". This introduces sociological reflections such as Joël de Rosnay's one on "ProAm". These are terms that refer to users that are simultaneously professionals and amateurs. They are also employed by Patrice Flichy [FLI 10] to describe his notion of amateur's coronation. We will come back to these problems later on. Axel Bruns [BRU 08] proposes instead the term "Produsage".

Model	Features	Limitations
DOI: Diffusion Of Innovation	Based on the observance of five rules	Mobile phones are successful even without abiding by these five rules
TRA: Theory of Reasoned Action	Mixture of attitude and product expectations via perceptions	Impossibility of knowing the future user's uses
TPB: Theory of Planned Behavior	Importance of the consumer's behavior	The consumer's behavior depends on the behavior of others
TAM: Technology Acceptance Model	Notion of perceived usefulness and ease-of-use	A large part of uses has no usefulness
UTAUT: Unified Theory of Acceptance and Use of Technology	Model that supposedly unifies the previous ones based on motivation and social knowledge	Set of all the previous limitations

Table 4.3. *Different adoption models*

In a spatial dimension, the emergence of a tool originates a process of adoption elucidated several times. Temporally, the diffusion of a tool is an effect that has been modeled, as if the consumer accepted the products after a certain amount of time and after being the object of marketing approaches. The whole matter is actually more complex. Consumers can refuse or accept the tool, whereby processes of diffusion ensue. The outright refusal sometimes takes the shape of total enthusiasm, and both postponed acceptance and misuse can be considered as approaches. Massive unpredicted usage alternates with applications without use. The same logic would be applied to treatment-related protocols. In both cases, it seems that doctors play an important role in the choice of a solution.

4.3. Some definitions

We will introduce some definitions that we will also employ further on.

DEFINITION 4.8.– *Habits are acquired dispositions that become second nature to us*[4].

DEFINITION 4.9.– *The forms of communication correspond to the set technical means that allow the diffusion of messages. It is a process: men can use machine to communicate.*

DEFINITION 4.10.– *Appropriation is made of two ideas. The former is the adjustment of something to an expected or defined use, to a target. The latter is the action that consists of making an object "owned" (as in "part of ourselves").*

All these notions coexist in the dynamics of innovation. This approach that separates habits, uses, memories and forms of communication is clearly a development of Jacques Perriault's proposition, contained in the introduction of his work. There he makes a distinction between "the project of usage, the instrument possessed and the function attributed to it" [PER 08, p. XI].

4 The definition used here could be qualified as "Bourdieusian".

Every new object upsets the habits of a set of individuals to different degrees. But what are *habitus*, habits, uses, memory? Memory and its forms of communication play a role in the relationship between the individual who gets treated or accepts prevention and the healthcare professionals. First of all, we should explain all of these notions.

Let us point out Bourdieu's line of thought. The *habitus* represents a structure that allows us to behave socially in relation to the class we belong to or we wish to belong to. It is a form of social conditioning, whereas habits are acquired dispositions that become second nature to us.

4.3.1. *The significance of the concept of habitus*

The notion of *habitus* developed by sociologist Pierre Bourdieu [BOU 79, BOU 81] seems appropriate to understand the different uses of medical technologies. Let us recall that this notion designates the set of dispositions and perceptions that an individual acquires through a process of socialization (education, social networking in a specific environment, etc.). By means of the *habitus*, certain kinds of "lifestyles" tend to form and to structure social groups into antagonistic interactions.

"The notion of *habitus* allows Bourdieu to combine the knowledge stemming from different traditions of sociology: the Marxist tradition, which reminds us how the forms of consciousness vary according to the conditions of existence, Weber's critique of Marxist materialism, which re-establishes the role of the social groups' world view as something related to their actions, and the *Durkheimian tradition, based on the study of the forms of classification such as it was adjusted by structuralism*" [SAP 04].

We deemed it interesting to compare, first of all, the dominant discourses about the three important aspects of ICT usage (diffusion of cultural productions, applications in the professional domain and political usage of the tool) with the social reality of these practices. The aim is to re-position telecommunications networks – including the Internet (and the relative discourses) – in a long-term historical continuum which allows us to appreciate the significance of the innovation represented by the new forms of communication, since this analysis is actually the analysis of a Kondratiev wave. We thus find ourselves in a new cycle of renewal of medical practices.

4.3.2. *The notion of appropriation in the health field*

Alaian Gras invokes a sociology of appropriation, whereas other authors seem to have banned the term "appropriation" from their vocabularies. He distinguishes three levels.

On an individual level, the user ensures that innovation suits his own personality. On a group level, appropriation refers to affiliations. We will see later on that these are linked to identities. They are as different as shared pastimes, age groups, geographical and social environments, and professions. Finally, on a cultural level, according to Alain Gras, appropriation is characterized by different uses. This three-level kind of approach is quite significant in order to understand the progress of healthcare systems.

4.3.3. *The notions of interaction and link in the health field*

The notions of interaction and link are central to our study. Interaction is the basis for the diagnosis that takes shape during the "singular discussion" between a doctor and his patient. It immediately refers to processes of socialization. Interaction is necessary for the process of "humanization" that we have previously mentioned in relation to Agamben [AGA 07]. If such unsophisticated medical technologies as a sphygmomanometer are becoming incredibly widespread, it is because absence is one of the deep forms of instability that men want to curb in every possible way [PER 08, p. 103]. We will see further on that this notion of instability leads us to that of simulacrum, which testifies to a form of religiousness associated with this humanization.

4.3.4. *Material interactions versus interpretational interactions*

Fombrun's works [FOM 98] teach us that the nature of interactions can be either material or interpretational [RIN 98]. This management-based approach can be scaled down to an individual level.

Thus, material interactions have to do with the resources and income – in the economic sense of the word we may associate them with – they drain. These resources are material, human and organizational. As for organizations, they consist of competences [PEN 59, PRA 90, BAR 91]. We will hypothesize that for humans, it is more a matter of knowledge and information. Doctors are required to study for a long time so that we can make sure of their knowledge acquisition.

Several researchers have hypothesized that other kinds of interactions exist besides the aforementioned ones. Weick classified them in relation to their interpretational nature. In organizational theory, they have to do with the managers' influences. However, we will also hypothesize that they affect human behavior [WEI 95]. These interpretational interactions are central to a doctor's way of working.

According to each of these approaches, action and practices are the result of different factors, as described in Table 4.3.

Kind of approach	*Expected interactions*
Classic or neo-institutional approach	From the institutional and temporal context
	From ideas, beliefs and values of the interpretative pattern
	From the organizational design
	From the intraorganizational dynamics
Weick's approach	From systems of beliefs, values and ideas
	From the interpretation of the environment
	From individual and collective interests

Table 4.4. *Differences between interaction-related approaches*

We will subsequently follow Weick's line of thought.

4.3.5. *Reputational interactions*

The theory of reputation is a kind of approach used as much for businesses as for individuals. In an article published in the newspaper *L'Expansion* in September 2001, Didier Pourquery interviewed Charles J. Fombrun and called his article "REPUTATION = MILLIONS". Fombrun, a professor at New York University's Stern School of Business, is the author of a classic book on the reputation of firms [FOM 96]. In 1999 he set up the "Reputation Institute", a center of study and research on business reputation.

He shows here how reputation contributes to value. His theories are found in every text concerned with the reputation of individuals as part of organizations. According to him, reputation is a source of value. Naturally the reputation level, unlike audience figures, is hard to assess with a simple figure. As for our media, reputation is a less precise marker which seems to be replacing this notion of audience. When the healthcare organization allows for choice, a doctor's reputation is as decisive a factor for the patient as the healthcare establishment he will go to.

If we follow Charles Fombrun's line of thought, reputation is not an entity of itself: it is always relative to a kind of reality. As far as organizations are concerned, reputation only exists within the relationships created by them and can be broken down into six major criteria:

– emotional appeal of an organization;

– products and services it diffuses, offers and sells;

– social responsibility;

– work environment offered to its employees (visible to the client);

– financial outcomes;

– these managers' views and leadership.

Many private healthcare establishments have understood the importance played by the quality of their premises while trying to make themselves acknowledged, and how it is at least as important as the other forms of quality. This does not seem half as true in Europe, where hospitalization is "managed" by the choices made by the ED and depends on the gravity of someone's condition.

According to this author, it is possible to apply this approach to individuals:

– emotional appeal;

– social exchanges;

– social position;

– life environment;

– economic performance, which can be assessed in relation to ones salary or his professional and social category;

– aura, charisma and leadership.

First of all, the notion of reputation is largely based on the relationship formed between the patient and the medical profession. Then, the notion of reputation and its corollary, represented by the audience, crop up as well as soon as we start talking about the media on the whole or people seen through the lens of the media. The medical profession is unwilling to accept that we refer to these notions, but any professor talking about vaccines, cancers or diets falls within this logic.

As for the health industry, several countries do not allow members of the medical profession to attract publicity or be too exposed to the media. This, however, often takes place indirectly. Local newspapers advertise the implementation of a new robot in a certain establishment or of a new surgical procedure in the other, or highlight how this or that doctor or surgeon has obtained research funding. Thus, reputations are built.

"Coaches" whose aim is to improve people's reputation have taken hold and are setting up plans for them to meet this objective. It may be a sign of the times but, in any case, these are elements of a society where reputation is held in more esteem than facts. What are the consequences for the reputation of a doctor, for whom any form of publicity or rating is forbidden by the Medical Council? We have to point out, on the other hand, that there are now in several countries rankings of hospitals and clinics published by the press and that medical data relative to the quality and appeal of these establishments is also of public domain. It is good form for a hospital to have automated surgical or scanning machinery available, even if it is very rarely used, and to have its installation as well as its first use on a patient published in the local press. This is considered common practice.

Further on we will come back to this notion of "reputation" as a source of trust.

5

The Dynamics of Diffusion

The dynamics of diffusion of medical innovations as well as those regulating the development of medical industrialization are topics that we must analyze in this book. At this stage, we should make two remarks. First, we have to notice that the effects of degeneration are hard to distinguish from those of these new long-term diseases. Secondly, we should point out the phenomenon of domino-effect adoption of solutions. Among these kinds of degenerations, those that have to do with sensory organs are the first ones to be affected by the development of technology. Our sight is treated, thanks to a laser slicing our eyes. Deafness is reduced by devices and cochlear implants. As a consequence, the medicine of the future, which will be industrialized, will have to improve our lives both quantitatively and qualitatively.

5.1. The basis for the dynamics of industrialization

The industrialization of health is highlighted by the possibility of using quantitative medical data in a field, where the subjectivity of the relationship between the patient and the medical staff is the norm. We can add to this the increasingly dominant need to feel safe and the desire to reduce the complexity of treatment.

5.1.1. *An approach – both qualitative and quantitative*

This approach was used in the past in the medical profession with the support of public authorities. The deadliest of epidemics were curbed by "quarantining". Strict hygiene rules allowed, by being rapidly adopted one after the other, to fight cholera. These approaches are not very compatible with the liberal expectations of medicine.

5.1.2. *Safety in the health industry*

The concept of health safety may be linked to the highest quality standards we can expect from the industrialization of health. When it comes to health safety, we can reason in terms of check lists, which is actually the case nowadays. We can draw an analogy with the notion of food safety, which was introduced while this sector was being industrialized.

Health safety consists of four elements: availability, access, stability and technical quality.

Availability refers to the notions of production capacity and related necessary equipment, and to the possibility of stocking medication or medical devices. This concept is not limited to hospitals. Let us take the example of home-care peritoneal dialysis. In this case it is assumed that the *santacteur*, who is treating himself at home, has enough room at home to store the different bags containing the products necessary for his treatment.

Access depends on the infrastructure available as much in the public as in the private hospital domain. It equally relies on the access to healthcare professionals and on their own access to patients. It also has to do with such factors as traveling distance or time. An increasingly important role is being played by the purchasing power available and/or the will to invest in one's own health.

Stability has to do with the stability of infrastructures, policies, management and procedures. As for healthcare, we have to point out that this desire for stability clashes with the impression of constant technical progress that characterizes this industry.

Technical quality is analyzed in terms of toxicity (mainly of medications), safety (of medications and devices), hygiene and sterilization.

5.1.3. *Health as a complex system*

As we have previously stated, life is an adaptable complex system. What characterizes all such systems is a constant evolution regulated by two kinds of processes: inertia or pseudo-stability and chaos. Let us recall that when an individual seems motionless, in medical terms a sawtooth tetanic process is taking place and the micromovements are therefore Brownian. Consequently, in order to measure someone's motion we will have to exclude these movements, which are of no interest to our analysis.

The complex and adaptable health system requires us to agree quickly on the topic discussed: the cost of saving a life varies dramatically in relation to whether we have to do with an accident, a disease, prevention, risk or treatment (we have willingly arranged the terms in alphabetical order). As we have pointed out, we have to add to this price the difference between individual and collective vision.

The shift from hemodialysis to transplants for renal diseases is typical of the technological changes that deviate from the system. Peritoneal dialysis is another development that makes the lives of sick people easier, challenging however the role played by institutions (in this case hemodialysis centers). In the future, artificial kidneys will certainly be available to us.

5.2. The advent of new concepts

We have pointed out one after the other the emergence of notions that are becoming more and more significant. At first, it was anticipated uses and actions that gave rise to behaviors before the phenomenon of bandwagon adoption took place. Concurrently, "the exacerbation of old peripheral functions" preceded the appearance of misapplications. This kind of approach is in polar opposition to the one proposed by Jacques Perriault. Uses result in fact from the creation of new practices. "I noticed several practices that deviated from the standard use, which were not mere handling mistakes". Jacques Perriault had sensed the emergence of new forms, since he writes a few lines further down: "They actually correspond to intentions, even premeditations" ([PER 08], p. 13). What this author forgot about was the co-constructed character that shapes the relationship between users and the technical system of what we will call "anticipated uses".

5.2.1. *Anticipated uses*

The anticipated use of technical equipment, whether medical or home-care, mainly pertains to the construction (or organization) of this network society. There, the notion of distance becomes less central and interpersonal relationships take on a new significance. We should also integrate two concepts developed by Bolstanki and Chiapello [BOL 99]: no one can make sure that the relationships established are reliable and, moreover, natural communication – the "face to face" of proximity – is "distorted" by the use of equipment ([BOL 99], p. 201). In other words, the tools used for telemedicine can only alter the relationship established between patient and doctor.

What's striking in relation to the observation of these anticipated uses is the actors' co-construction of the media arguments. Some go as far as claiming the existence of the co-construction of a business, hospital or family culture. Using telemedicine isolates the individual at first. "Dissemination of these tools in daily life does not produce an isolated individual oblivious to social constraints, nor has it triggered a process of globalization of standardized social practices" ([CAR 05], p. 17). However, we should notice that this discursive approach involves some risks, owing to the gap between words and reality.

These two authors even mention the "production of the social culture", which is carried out in two ways. The first one has to do with how these tools "make us communicate", even in those circumstances in which we formerly would not[1], whereas the second one pertains to how communication creates culture since these acts of communication "are texts" in themselves. Thus, Caron and Corinia take up Gergen's theories on the "absent presence" ([CAR 05], p. 45; [GER 02]).

5.2.2. *Bandwagon adoption*

This approach has been widely used by mobile network operators all over the world by favoring calls on their own networks. To allow a free call

1 We will delve into this matter in section 5.2.2.2 on misuses.

between A and B will lead A to advise B to buy the same mobile. B will use his own mobile to call other people, and the fact that the communication between A and B was free will be largely compensated for. Besides, the high costs involved in making calls from a landline to a mobile phone have contributed to the purchases of mobiles. There is no example in the medical sector that can be as clear, but if we think of people advising their good friends to buy a certain medication, we can come close to it.

5.2.2.1. *Old peripheral functions*

Few marketers know the actual reason why this or that kind of equipment has become successful. Our analysis will only present one of the causes of this success, which we will call[2] "old peripheral functions" (OPF).

It may be interesting at this stage to pay attention to the production of these OPFs. Two approaches have been endorsed by manufacturers. The first one consists in introducing an internal function via an object that is more complex than needed, and waiting for misuses to develop[3]. The second one consists, on the contrary, in promoting an "ecosystem" in which a multitude of actors propose solutions.

5.2.2.2. *Misuses*

This approach is a development of the one proposed by Jacques Perriault in relation to directions for use. It is also in opposition to the merged approach he proposed in his work on imbalances. Anne-Marie Laulan was one of the first to describe some misuses. Her work fitted into a study on opposition to the information system that "violated" users. She also discussed the notions of deviance, variation, misapplication and arpeggio [LAU 85]. Current scandals are replete with the misuse of medications, the effects of which are not always positive.

5.2.2.3. *Uses of the first kind*

This usage is represented by the intensification of a function of the product. In our surveys, we have discovered that consumers are more

2 We should not forget that terminal subsidy is an important variable. It explains the fortune of many firms and especially Nokia, which benefited from the better level of equipment grants provided to operators, simply because they could personalize mobiles more easily.
3 As for mobile phones, it was the "watch" function.

interested in functions as soon as they deviate from the basic function of the product. In other words, the use of a medical device or protocol of treatment exceeds the one for which it was intended.

5.2.2.4. *Uses of the second kind*

These uses appear in those situations that we can qualify as "no win, no win".

In this kind of circumstances, it is essentially a matter of the meaning given to "non-time" and to "non-places", if we follow the terms proposed by Marc Augé [AUG 92]. These places have no known cultural meaning and are not linked to a personal association or agreement between communicators. They are banalized or banalizable: bus stops, stations and airports. These places have become treatment sites, for example for a diabetic who tests and injects him/herself with insulin.

5.2.2.5. *Uses of the third kind*

A use of the third kind is linked to the appearance of bans or ethical rules. Knowing about usage at first leads to operational or normative consequences in relation to those involved in a sector. As a precaution, usage will be most often intensified. The role played by abortion in countries that neighbor nations where this practice is prohibited by law perfectly illustrate this point.

Uses	Operational consequences
Anticipated uses	Offer of suitable pricing system as soon as uses are identified
Bandwagon adoption	To favor equipment changes in order for them to take place
Old peripheral functions	Integration into the proposed equipment or services, even if the link with mobile phones is weak
Misuses	Presentation of misuses in documents for information purposes, in the hope that they will be circulated

Table 5.1. *Operational consequences resulting from the knowledge of uses*

The question boils down then to understanding whether users will be responsive or not to these approaches.

5.3. Attempting to reduce complexity

Attempting maneuvers which are often ineffective, the opposition to the logic of action or the shift from the actual passing of time to asynchronous dimensions can all bring about this reduction. Finally, another approach consists in establishing rules about the correct use.

5.3.1. *A three-dimensional model to conceive products*

Real time, which was the historical dimension of the first tools used in medicine – such as the telephone – has given way to asynchronous forms of communication. This is a utilitarian logic, based on effectiveness. The possibility of asynchronous communication is the source of a communalist character, which is the element of a logic of adhesion or integration. The community takes shape by accepting its future members within an asynchronous means of communication. This acceptance introduces a logic that some regard as critical: the consequence of men's necessary autonomy, which we analyzed in the previous chapters. This approach was also defined by three kinds of logic of action [JAU 11].

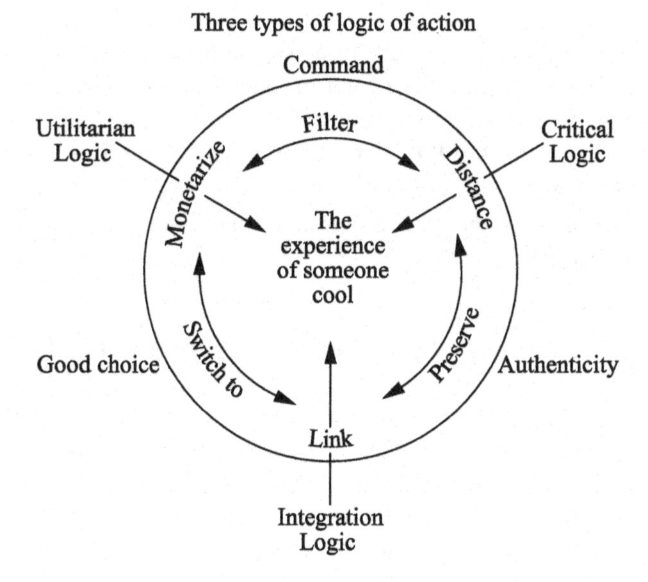

Figure 5.1. *The three types of logic of action proposed by Jauréguiberry and Proulx*

5.3.2. *Real time versus asynchrony*

The actual time of a mobile connection corresponds to a spaceless and instantaneous kind of time. The connection speed tends to eliminate the time frames necessary to cover great distances. In other words, less and less distance needs to be covered in order to establish a connection. On the other hand, there are more connections to be established. This notion of real time gives rise to a notion that is engrained in our minds: mobile phones represent saved, or economized, time. When these technologies were being launched, this was the providers' favorite selling point. It even made it possible to calculate returns on investment, which were often misleading. This notion somehow corresponds in the medical sector to the creation of stroke centers.

According to Fisher [FIS 01], speed is a cardinal virtue. It considers respite something that verges on laziness and leads us to sacrifice the time we would dedicate to reflection in favor of fast communication and compulsory answering ([FIS 01], p. 79). As Rifkinks [RIF 00] underlines, access – or, in other terms, the relationship – has to be possible in order for this to hold true. Hence, the emphasis put on the gateways to this new form of long-distance medicine: stroke telemedicine.

Asynchrony is therefore a reaction to the tyranny of real time. Being asynchronous makes it possible to get organized so that we can process the information to which the individual reacts. This concept, proposed by Haffner and Lyon [HAF 99], has been verified by our analyses. The instantaneous aspect of communication has vanished. This kind of technology belongs to the phenomenon of bandwagon adoption we have previously described.

5.3.3. *The assessment of good use in the medical field*

The assessment of "good use" in the medical field will be made through indicators. The UK set up an indicator called NICE (National Institute for Clinical Excellence), which aims to ensure equality in terms of access and healthcare standards. Its policy is to obtain and optimize healthcare spending while trying to find the most economical performance. In order to do this, it employs an indicator called QALY (Quality Adjusted Life Years), which is virtually the same as the aforementioned indicator named HALE and represents the number of years spent in good health. It is very useful for degenerative diseases such as Alzheimer's.

The problem posed by the assessment of quality of life lies in how it can be gauged in several ways, for instance by a disease specialist who diagnoses a patient and forms an opinion on the profit made. The "standard gamble" approach is another solution. It represents the risk of dying, whether because of surgery or medications, that patients deem acceptable to get out of their state of ill health. This risk varies according to the patient's knowledge. Besides, a *santacteur* can see this risk in a very different ways and may even be risk averse. The third method is called Time Trade Off (TTO) and consists in confronting a patient with a choice between a better quality of life, with a reduced life expectancy, or the same health state until the end.

A forth kind of approach is in line with the logic of palliative care and consists in assessing the patient's physical and mental suffering by means of psychometric tests and mechanical analyses of behavior.

We should not forget that quality of life also largely depends on the social environment and on the individual's financial resources. These approaches, however, do not take into account the effects on family caregivers, especially the patient's support. Finally, such actions cost in terms of public policies and we should therefore ask ourselves to what extent the individuals themselves will have to pay. In this case, the situations could be quite uneven.

5.3.4. *More than a blend of engineering science and traditional medicine*

Without getting involved in the debate about science and technology, we should admit that the industrialization of medicine does not consist in introducing engineering science into the discipline of medicine itself. It goes beyond this mere process. To make things simple, we can say that engineering science belongs to the dimensions of engineering. It deals with complicated systems. The essence of engineering is the ability to master "a" situation. However, the industrialization of medicine takes place within a complex system. Thus, there are two possible solutions.

The first consists in drawing out some sub-systems that can be solved with the application of engineering science, which is currently the case for medical imaging.

The second solution, simultaneously more complex and more rewarding, consists in finding out the elements that make up the complexity of life.

Besides these two reasonable solutions, a series of large-scale European programs, aiming to modelize life, have emerged.

6

Digital Identity and Health

The notion of identity is simultaneously related to different forms: sameness, similitude, belonging to a group or community, uniqueness, civil status, personal status, etc. Sometimes, identity refers to a kind of permanence (in the sense of continuance) or to geography. We talk about IDs without distinguishing between people and companies, even if we intuitively understand that they are two different things. Let us point out that identity is also the relationship between two equal terms (mathematically speaking). With regards to the medical field, the identity question is not pointless. In the mass production of identical cars, the identity of each corresponds to its license plate. Evidently, this kind of analogy cannot be reproduced in the health industry.

6.1. Digital identity

If we consider our digital tools, it is their own specific nature that leads us to think about these elements:

– defining the term digital identity;

– defending its anonymity on the network (which is performed through technological solutions);

– understanding the concept of digital identity;

– pointing out the specific nature of digital identity in relation to the mobile phone dimension.

Service providers have come up with some solutions. In the beginning, they consisted of setting up services that pool users (Passport, Liberty Alliance, Enum, OpenID, etc.). Digital identity is often limited to an electronic signature and some cards. This approach is in line with a strong component of electronic management or administration. Identity has been at the center of the debate on the different set-ups of personal records, to the point of losing its usefulness. We should add to this the legal problem related to the ownership of personal data, which is one of the most complex. For clarity, we will willingly turn a blind eye on these problems.

6.1.1. *Identity*

Psychological definition: identity is the representation of my relationship to others, of my presentation to others, and of my representation of others' representations.

In our digital society, identity is defined by the environment, our appearances, habits and habitele. There is clearly a strong link between senior citizens and their identity. The environment has to be changed in order for them to lead a decent life. However, appearances, habits and habiteles also need changing if we want digital healthcare systems to be successful.

"Habitele is used to label the various distant connections with various social worlds that we are able to handle by carrying devices (phone, credit cards, IDs, keys and access cards) and traces that keep us in touch with these worlds". The term habitele was proposed by Dominique Boulier (Media Lab, Science Po Paris) and Artur Hecker (Telecom Paris Tech), from whom we have drawn this excerpt. According to Charles Tijus (CHArt Laboratory, University of Paris 8), who follows their line of thought, the question of senior citizens' access to digital technology has to do with understanding the elements that will strengthen their identity and self-representation.

We are just at the beginning of this digital revolution that cannot but continue to grow.

6.1.2. *A manifesto for the advent of the new man*

Soriano and Finkielkraut are interested in the advent of a new man made possible by electronic networks. “The network dissolves those identities and alienating communities – the jails of our conscience – in favor of blended, freely granted, and modifiable forms of identities and perfectly porous virtual communities, allowing us to eliminate in one go intolerance, intraspecific conflicts fuelled by difference, as well as the hell represented by the Other” [SOR 01].

6.1.3. *Likeness as a source of community*

The notion of community has been studied only recently and its forerunners are sociologists such as Durkheim and Weber. The concept was formalized by Tonnies [TON 87]. According to these intellectuals, it is a sort of reaction against the atomization of social life and, therefore, an organic natural will. It presents itself as a tripartite will: chosen will, artificial will and rational will. The necessary conditions for a community are elements inherent to the organic will, but it is rational will that produces society. We have to link this notion of community to that of habiteles, i.e. the networks that enable the existence of these very communities.

The study of communities should become necessary with the creation of the healthcare networks and hospital groups necessary to optimize the health offer.

Tribalism and communities involve a shift away from a narrow view of the individual. In this case, the individual exists only for and thanks to the social dimension. In order to identify the tribe, we should focus on the notion of restricted groups or look for customs, habits, traditions and rites. Starting with this first action, we should later identify the function played by the individual within the tribe. It is the role of thought leader, adviser, innovator, follower, or it corresponds to the creation of relationships of subordination or reliance.

6.1.4. *Nomadism, territory and health*

The individual is plunged into a consumer society in which we are currently shifting from mass consumerism to ultra-differentiated consumerism. Do we "consume" health in such different ways that it needs to become local or, in other terms, independent from national public health? This development would mean that any national health plan, such as those conceived for cancer or Alzheimer's, is doomed to fail. Consuming is often associated with the worship of individualism. Man becomes an emancipated subject and yet lacks any point of reference, a situation where even a renewed community could only be explained after a sustained effort of detection. The real question would therefore consist of knowing what a community is and what are the types of postmodern communities. This formulation revives the debate about the existence of communities rather than tribes.

6.1.5. *Community, network and network externality*

The information-centric society creates an "uncertain" society, which signifies the end of provincialism. A community based on interests ignores geographic borders. Some sociologists conceive a sort of "founding nomadism", created by the social link diffused by ICT. This social link would therefore become the cement for society.

The social link is then communicational and lineal. The traditional relationship between the social link and society turns into a dynamic dimension characterized by connectedness and community. This notion of connectedness should allow us to explain how habiteles are formed.

6.1.6. *Electronic community and mobile-phone reality*

Electronic communities seem to give rise to a redrawing of boundaries, performed no longer in geographical terms but based on a different context. The notion of interest is the first that comes to mind. That is why we proposed a double classification in a *Fondation Internet Nouvelle Génération* (FING) working group. The first approach would separate wholly virtual communities from those semi-virtual communities that have a real component (for example the fan club of a football team). The second approach would distinguish between instructive, affective and play communities.

6.1.7. *The importance of access*

Reachability and accessibility are two key notions in terms of new networks. This approach was the topic of a debate on the very terms of this reachability. It represents very different forms: perfect geographical coverage of the network, permanent connection to a global network, constant connection to organizations and companies and unique number. Often the definition of reachability has become radical. It is defined as the possibility of being reached on a mobile device anywhere and in the best possible conditions. In this case, reachability depends on the presence or lack of network, the quality or sensitivity of the device, its travel speed, its location, etc.

This notion of reachability becomes significant when we deal with different symptoms and diseases. A specialist's reachability and access time in the case of a cerebrovascular accident plays a crucial role in determining whether the person hit by stroke will survive. Notice how in this case and in general as well, we should distinguish between the time necessary to access the healthcare facility, where the thrombolysis will be carried out, and the reachability of the neurologist who will perform it and who may be absent.

6.1.8. *Civil status and personal status*

Civil and personal status are two forms of the same individual, namely two forms of his identity. Philippe d'Iribarne claims that a sense of identity can only be nourished by the conscience of each being. Now, the reference scene of a society is for the most part unconscious [DIR 08]. This does not prevent the creation of objective elements of identity, like someone's fingerprints, but then we are concerned with something very different from the elaboration of a subjective sense of identity.

Our civil status is defined by rules and statuses – for example single, married, divorced, etc. – that depend on a well-defined approach which is necessarily static. In other terms, information is relatively stable.

Someone's personal status is on the contrary a dynamic form. For instance: I am tired or in good shape, standing or lying down, etc.

The notion of civil status refers to that of nation. According to some people, it is presented as a mythical universe elaborated only recently in historical terms [THI 99]. As for France, it was Napoleon III who created the myth of the Gauls and Alésia. This concept seems to contradict the equalitarian principle of the French Revolution, which is why it has always been rejected by communism. According to some other people, civil status is often presented as a "principle of bourgeois ideology and policies that propagandizes isolation and exclusion". It is in line with the chauvinism of great countries or with another similar concept, the distrust of other countries. Formerly, it was used to justify the war between countries and colonialism. We should point out that not much attention has been paid to these notions of identity when applied to health in relation to the set-up of medical files in different healthcare systems.

6.2. The identity crisis of the information-centric society

Introduced by "globalization" and the impact of ICT, this identity crisis can be analyzed in terms of *mondialisation*. It consists of the end of ideological opposition, the conclusion of the Cold War, and the omnipresence of capitalism. Identity is currently struggling in an atmosphere – reduction of nationalism, increase in regionalism – characterized by the search for a single market (Europe, Latin America). Therefore, the crisis of "nation states" is contemporaneous with the emergence of new local and regional identities. We will not come back to this point. However, for similar reasons, healthcare systems are going to develop specific identities.

Several authors, like Philippe Queaux, mention the emergence of a civilization [QUE 98], whereas others are discussing an identity crisis. Finally, yet another group regards the Internet as Diderot's dream come true, i.e. a network-built immense library. Thierry de Montbrial recalls that "nation states are born from the ruins of civilization and medieval cosmology, and are the result of the invention of the printing press (...) Nowadays, our vocabulary is unfortunately unable to describe the permanent historical dynamics that drive us along".

The sources of identity are, first of all, criteria of communal allegiance, whereas anciently they would be based on ethnicity or religion. They were characterized by their coexisting durability and commonness. Modern sources tend to be linked to national identities and shared future projects which are simultaneously social, corporatist, and ideological. Recently, they have been clashing with the crisis of nation states. Post-modern sources are, on the other hand, freely adopted, ephemeral and disposable, as Soriano and Finkielkraut reveal in *Internet, the Troubling Ecstasy* [SOR 01].

6.2.1. *What role does identity play in the health industry?*

Only "scars" can make a body or a machine unique. Anthropology has provided marketing with a powerful idea. Consuming is not only utilitarian but also consists of creating meaning and exhibiting at once individual identity and social belonging [LAD 04, p. 29]. Hence the idea that the applications and functions we use on our mobiles betray in us a sense of belonging and, on the other side, the desire of some actors to harness them. Thus, an important form of identity in the health industry consists of the sense of belonging to a group of patients or the fact of suffering from a certain disease or disability.

Perriault makes an important remark. "Pierre Schaeffer, in a recent conference held at the Collège de France, when discussing the etymology of the term 'communication', underlined how the association of the prefix -com with the root -munus, a statutory job (like a magistrate), shifts the etymology toward the meaning of 'identity exchange'" [PER 08, p. 55]. The user of our current devices actually does more than merely exchange his identity; he also exchanges his sense of belonging. Moreover, identity, as much as belonging, is dynamical and consequently does not make much progress over time.

DEFINITION 6.1.– *Socialization is defined as the set of rules that the members of a given society must learn, together with the norms and values particular to the society.*

This process of learning is first conducted through interpersonal communication. A part of these rules and norms may be explicit, whereas another may not.

Many authors have worked together to define the stages of this socialization, which establishes moments of change in the economic dimension. This socialization is a constant process for the individual. These stages have been qualified as human needs.

The 14 human needs

To breathe

Someone's ability to maintain an adequate level of gas exchanges and good oxygenation.

To drink and to eat

Someone's ability to drink or eat, chew, swallow, be hungry and absorb enough nutrients to accumulate the energy necessary to his or her activity.

To eliminate

Someone's ability to autonomously eliminate excrement and urine, ensure his or her intimate hygiene and eliminate body wastes.

To move, maintain a desirable posture and keep adequate blood circulation.

Someone's ability to move independently, or thanks to mechanical means, manage his or her home adequately, feel somewhat comfortable and know the limitations of his or her body.

To sleep and rest

Someone's ability to sleep and feel well rested, and to manage tiredness and energy potential.

To get dressed and undressed

Someone's ability to put on, take off and buy clothes, and to create his or her own physical and mental identity.

To keep one's body temperature within the normal range (37, 2°C)

Someone's ability to adjust clothing in relation to his or her environment and appreciate its limits.

To be clean and well-groomed, and to protect one's integuments

Someone's ability to wash, keep an appropriate level of hygiene, take care of him/herself, use skin products, feel a sense of well-being, feel comfortable in his or her own skin and look at him/herself through other people's eyes.

To avoid dangers

Someone's ability to maintain and promote physical and mental integrity by being aware of the potential dangers present around him or her.

To communicate with others

Someone's ability to be understood and to understand, thanks to behavior, the spoken word or a code, to integrate in a social group and live fully his or her emotional attachments and sexuality.

To act according to one's beliefs and values

Someone's ability to know and promote his or her own principles, beliefs and values, to make use of them in relation to how someone desires to live.

To work in such a way that there is a sense of accomplishment
Someone's ability to engage in play or creative activities, pleasures, make use of them for his or her self-realization, maintain self-esteem and keep a role in a social organization.
To play and participate in forms of recreation
Someone's ability to relax, educate him/herself, get involved in activities not related to personal problems and feel personal satisfaction from it.
To learn
Someone's ability to learn from others or from an event, evolve, be adaptable to change, become resilient and pass on knowledge.
Source: [HEN 78]

Box 6.1. *Human needs*

Dependence is a part of identity.

6.2.2. *Dependence*

The notion of dependent people was defined by the Council of Europe in 1997.

DEFINITION 6.2.– *Dependent people are individuals who, because of physical, psychological or intellectual deficiencies, need assistance and/or significant aid in order to go about their everyday activities.*

Thus, the term dependence designates concretely the kind of help needed by the elderly (60+) in order to carry out certain basic daily activities, as well as by mentally or physically disabled people who cannot be self-sufficient.

Some philosophical problems arise at this stage. Should we attempt to help at all costs or should we, on the contrary, try to make people more autonomous? Do we turn into debilitated beings as soon as we make use of gerontechnology?

6.2.3. *Internet identity, class, community and tribe*

Those digital identities that show our membership to classes, communities, or tribes go well beyond the explication[1], proposed by Renaud Camus, based on a central "bourgeois" model. However, we are obliged to go back to Marxism when we highlight class conflicts. As these notions of class, community and tribe will be we largely used later on in this book, we will tackle them rapidly in order to illustrate their limitations.

The notion of class derives from Marxist philosophy. Nowadays, it is being challenged by sociologists.

"The officialization of the notion of class carried out by the state raises two problems in particular. The first pertains to the double nature of the concept, which is simultaneously descriptive (it must allow us to classify and describe systems of differences) and normative. These differences rely on uneven distribution and a society with egalitarian ideals must set its sights on their disappearance or decrease. The second problem has to do with the positive or negative character of the description and the difficulties involved in creating a unique stable descriptive language for the discussion of upper classes and popular classes" [BOL 08, p. 127].

Boltanski points out that we have evolved. The ideology that was dominant when his previous work was published was based on intelligence. Intelligence was validated by one establishment: education. The best individuals were therefore part of "great institutions". According to him, "we can characterize the new concept of domination through the paramount role that it assigns to the 'government by norms"[2]. Therefore, this notion leads us back to questions like the digital divide, insofar as we regard owning mobile phones – and perhaps in the near future a device for the assessment of our well-being and good health condition – as the norm.

When recalling Olivier Galiber's thesis [GAL 03], we studied in detail the notion of community by starting from Tonnies' approach [TON 87]. This PhD dealt with the consequences of the marketing exploitation of norms and the forms of online community socialization (virtual communities). In that thesis, we defined the notion of ICSA (Internet Community Services and Applications). Now I think that this notion should be replaced by that of

1 We refer mainly to his last work: *La Grande Déculturation* [CAM 08].

2 This idea is particularly supported by Thévenot at the EHESS.

device, which we discuss here. Later on we will come back to the importance of this last notion.

The notion of the tribe was widely debated in my circle between 1981 and 2011. Most often defined as a social formation created after a state or certain states are built, it was criticized by ethnologists. Some of them would like to limit its use to parental relationships. The criticism put forward by sociologists too often emphasized the ways of separating the good from the bad. Political scientists saw in tribes the reminiscence of colonialism. Management professionals continued to regard them as subsets of the sections on which managerial actions focus. More recently, social apps propose to group our friends into "circles", which correspond to social formations. This is why I ask the reader not to discard these notions at first. Doctors have often confused this notion of tribe with groups of patients suffering from the same disease, since the mode of transmission of an external agent was crucial in relation to the disease.

6.3. Cards as new kinds of identities

The health industry is ruled by detailed and many-sided regulations that must be respected by any process of industrialization. This set of rules is based on identifying the actors involved.

6.3.1. *Developments in relation to the awareness of a sick person*

As we remarked in another work with Marie Marsaux: "During their unique discussion, both the doctor and his patient have rights and duties. The latter, whether simply ill or hospitalized, enjoys several rights that derive from a slow type of progress that started at the end of WW2. A sick person was considered as someone who was being "assisted", then as an "object of treatment", and from 1974 onwards as a "subject of treatment", who enjoys the rights of a person who is developing while part of the health industry. The Law No. 2002-303 of 4 March 2002, concerning patients' rights and the quality of the healthcare system, turns the patient, called a "user", into an actual "player" within the health system, who for example participates to general health assemblies and defines the progress that needs to be made in the health industry both individually and collectively. This is how the "health democracy" is born. The system has been in a perpetual state of evolution ever since this law passed. The Laws of 6 and 9 August 2004, about bioethics and public health policies, and the Law of 22 April 2005 on patients' rights and the end of life, are particularly significant.

Those who play a part in a medical action must be identified. Currently, the patient is identified throughout the care process thanks to his "*Carte Vitale*" (French national insurance card). The healthcare professional himself is identified by means of a "*Carte de Professionnel de Santé*" (healthcare staff card). Therefore, it is cards that display the patient's identity.

6.3.2. *ID versus healthcare cards*

In several countries, ID can be used by patients. Other countries, such as France, preferred to separate ID from healthcare cards, which then become "rights cards". Therefore, they are not only used for identification purposes, but also to find out about patients' rights.

6.3.3. *The French "Carte Vitale"*

Every French citizen above 16 is identified nationally by means of a unique National Identification Repertory number (NIR) and is provided with a "*Carte Vitale*". This card was created in 1984 and circulated nationally in 1998. It includes the holder's NIR as well as an INS-A (*Identifiant National de Santé* or Health National Identification), a randomly chosen number made up of 12 characters. The INS-C, another personal ID number made up of 22 characters, can be obtained by a deterministic calculation (mathematical or algorithmic) based on the information read on a patient's "*Carte Vitale*". These INS ID numbers are not classified information; however, they remain a part of the individual's privacy.

6.3.4. *The healthcare staff card*

The healthcare staff card is a profession kind of ID which is protected by a secret PIN. It enables us to verify the holder, besides proving its professional certifications and conditions of practice. It can also be used as a kind of electronic signature. There is however a small difference: in France the healthcare staff card can be obtained free of charge by any healthcare professional as defined by the public health code but under similar conditions in other countries, such as Germany. Using this card becomes necessary to access or convey any piece of medical information pertaining to the private sphere.

Currently, these cards only provide ID data and administrative information. In many countries, especially in France, the choice was made not to include any health-related data.

6.4. The genetic revolution as an ID carrier

The genetic revolution is a key point in relation to the new kind of health identity.

6.4.1. *Ease of implementation*

The first mapping of the human genome cost several billion euros the exact figure is not known. Nowadays, modern methods will allow us to sequence a whole genome for less than 100 dollars, provided the test machine is available. For diseases like cancer, the part of the genome that may be affected is currently being sequenced. As a result, we can quickly imagine how a map of our genome may be included in health cards like the French "*Carte Vitale*".

6.4.2. *Additional data in the health industry*

It is therefore important to add environmental data to this genetic map. For diseases like cancer, diabetes and cardiovascular problems, interactions with the environment play a significant role.

We should also include family predisposition as well as lifestyle, particularly in terms of diet. Therefore, adding this kind of data to patients' files should make medicine more prognostic.

6.4.3. *DNA, a source of identity*

Sixty years after its discovery, DNA has become an essential factor in terms of identification. It is used as much in the health industry as in forensic science. This therefore raises social questions. Individual freedom, security and the potential management of populations by means of recording are all risks that need to be compared with DNA use, which is necessary for the advent of a new kind of personalized medicine. As a result, the economic, political and social issues involved are significant.

The other issue has to do with a new type of medicine that stems from the *santacteur*. Since he wants to be the master of his health, he requires new forms of medicine.

In a piece called "My Medical Choice", featured on the opinion pages of the New York Times, American actress Angelina Jolie claims she has had a double preventive mastectomy. According to her, it is a brave choice. She has revealed she made this choice in order to prevent a significantly high risk of developing breast and ovarian cancer due to a faulty gene. The gene she mentions is called BRCA1. In the article she states that in this case the risk of developing breast cancer is 87%, whereas the figure for ovarian cancer is 50%+. The risk of cancer is thus reduced to 5%. This news has been featured by the press all over the world.

7

Access to Information, Access to Health

The notion of access is fundamental to telemedicine and e-health. It could be broken down into access to medicine and access to the network, if the two were not linked.

According to Anthony Rifkin (Foundation of Economic Trends, Washington, USA), digital economy is not a new kind of economy, but actually consists of the introduction of new forms of commerce and unprecedented situations. He regards digital economy as relying on fundamental concepts represented by a revolution in terms of access. The same can be said about the health industry: it is access to health that has increased our lifespan. It goes without saying that the lack of access to health in certain countries is the factor that explains lower life expectancy figures.

7.1. The access revolution

Anthony Rifkin points out, first of all, how the success of digital economy depends on an access revolution, which allowed us to exploit resources that had up to that point remained inconceivable. This is also the case for the implications of digital technology in the health industry.

7.1.1. *More resources, more trading*

We therefore have new resources that intensify trade.

He compares the effects of this revolution in terms of speed, resulting from connectivity, to the changes brought about by electricity at the time of the Industrial Revolution.

Moving towards the "cyberspace" means going from marketplace to marketplace, which is not a new phenomenon, since bankers were already trying to do this in the 13th Century. The difference lies in that, in our current economy, time is becoming more and more financially valuable.

His second point is the intensification of the trading economy in the face of a point is the traditional market that is just too slow.

The role played by stocks was paramount in the previous kind of economy, but now the notion of the flow is prioritized. Thus, we have shifted from a production-centric industry to a production of culture and content. Anthony Rifkin explains that certain companies of the new economy are better represented on the stock market than groups like General Motors (one of the largest concentrations of goods and machines in existence), despite having very low levels of concentrated physical capital.

7.1.2. *The impact of networks: disappearance or concentration of actors?*

"Are these old companies going to disappear?" Anthony Rifkin replies that they will not, if they migrate toward the new types of economy. For instance, car manufacturers should stop selling traditional cars and instead focus their production on "connected" cars, which can actually be used as a means of transport. All sectors of the economy are affected.

Networks concentrate markets and, consequently, powers. Antitrust laws will therefore have to be redrafted. There are good and bad networks: we will have to keep the good ones and eliminate the bad ones. Time is limited and each experience is made up of things that we have paid for. We lead contract lives. Relationships are becoming more and more commercial and less and less social. The most horrible nightmare we may conceive is the selling and buying of time. According to some, the current revolution allows us to save time, but we are actually running after time. As a consequence, we may wonder, where is this revolution? Is it not a step backwards instead?

7.1.3. *What society?*

Let us consider the example of emails and GSM. Does being constantly bothered constitute progress? How can we really use this technological progress to our advantage? It is a matter of attributing value to time. Old English dictionaries defined time as a metaphysical condition, i.e. what God made of the machine. As for love, we will be effective and make the most out of our feeling in the shortest time possible and with the least amount of effort. We have added the notion of "sufficiency": the machine has to be at our side.

Cultural diversity, as well as biodiversity, is a significant element. Pipelines are goods on which we can make a profit only if they contain something. We can say the same about the telecommunications industry. The wealth of cultural diversity must be preserved. This is only one of the many battles that need to be fought.

Another point consists of preserving trade culture. Everyone uses new technologies to do business. Therein lies the difference between the inherent value of culture and the internal value of trade. The battle against GMOs (genetically modified organisms) is the same as the one fought in favor of cultural diversity: the ways in which our food is produced constitutes a cultural site. What should we think, therefore, of a genetically modified man? Local culture is as powerful as global culture.

Trade was born from culture and not vice versa. Thus, we have to establish a state of equilibrium between the two. The World Trade Organization (WTO) is too narrow-minded an institution. Culture is an art and anything not related to commercial contracts. If it had to collapse, how long would it take us to build it up again? Medicine is also a kind of art that is being industrialized, as is the case nowadays.

Who can fully understand the local aspect, who is going to fill the void left by culture? A fourth kind of force: mafias. There are antidotes to *mondialisation*. Fascist and religious groups are convinced that only their own type of culture is worthy or, on the contrary, that other cultures should be prohibited. The ways in which medicine is practiced are certainly being affected by these cultural phenomena. It is, therefore, a matter of culture as a counterweight to *mondialisation*. In other terms, will health be industrialized through processes typical of *mondialisation* or will it, on the contrary, be based on local medical practices? From what we can gather, the "*choc*" would be more violent in the first case.

One of the mainstays of this new society is the explosion of data. The Web is made of data, so how can we turn it into useful knowledge and wisdom? This is the problem we are faced with. How can we make these experiences interact with our lives?

7.1.4. *The two faces of networks in our society*

This leads us back – very concisely – to the notions of human performance and technological progress.

"I favor human and social performances over technological ones". The apparent facility of technological progress is countered by the opposition and difficulties involved in the communication between men and society. This theory is especially sustained by Dominique Wolton (Research director at the FNCSR).

With the advent of the Internet, which is still a booming market, we can notice a sort of intellectual short circuit. There is no link between technological progress and the development of society. "The Internet, a tool for optimizing the economy; the Internet, global political utopia". This utopia affects quite heavily healthcare practices and, naturally, the exchange of information within the health industry. This is the reason why we need control systems: if we want democracy, we have to make sure we are in charge of it. The difficulty will lie in aligning different forms of regulation, for instance the code of ethics vital to the health industry with the management of free information Internet practices.

If we followed on with this simplistic approach we would find as a result a reduction in terms of human communication and an increase in media coverage. One of the key elements of this schizophrenic evolution is the exchange movement that we can find in the unique discussion between patient and doctor and the mediatization of his health condition on social networks, done in the hope of finding the ideal treatment. The same individual will demand security and confidentiality in the first circumstance, but he will then exchange them on social networks for responsiveness and an answer to his health problem.

There is a gap between the extreme facility that characterizes the Internet and the drop in communication. Exchanging messages does not constitute communication. Everybody is identified by the term network, when it is actually an interconnection that restrains our freedom. According to Wolton, Internet Utopians anticipate revolutions, but we need to clearly define what is at stake in terms of politics.

When everyone can access the Internet, understanding will no longer be a reality. The *mondialisation* of networks consists of exchanges of all kinds. Technological progress will highlight that a lot of effort is needed in order to understand one another. Therefore, it is necessary to introduce a humanist sort of progress to prevent a potential form of connection from creating friction.

7.1.5. *Towards mutual tolerance*

The Internet is a tool that restructures capitalism or, more generally, society, but it does not constitute a form of progress toward more democracy. The Internet provides us with a certain kind of technology. As for economy, we need a political project, which is a real challenge for Europe. The Internet will also play a role in re-structuring the healthcare offer.

We should implement another political project besides the creation of networks. Access does not necessarily entail understanding, just as hospitals do not ensure our survival.

The core of communication is mutual tolerance and the positive effect consists of the organization of a form of coexistence with those who do not agree with each other. The more technology invades our lives, the more we will have to escape from it. Men are not mere information systems; they are made of contacts. They have to tolerate one another. In healthcare establishments, it is the communication between doctors and the other medical staff, especially nurses, that is at stake.

7.1.6. *A theory of access equality*

The Internet and telephones are two different things. When on the phone, people talk and, therefore, they can make themselves understood. When it comes to the Internet, there is a demagogical aspect to the matter that consists of trusting the communication. Being able to access a large pool of information is important, but it does not guarantee the acquisition of knowledge. The greater the number of keyboard-centric mediations, the greater the need for mediators that can process this information and truly gain access to knowledge. The access to knowledge, which is one of the significant contributions of communication networks, will have a major impact on the modes of treatment.

7.1.7. *A modified autonomy level*

The level of human autonomy within society is substantially altered by the advent of the Internet. Man is finding a new place in the world. New technologies will modify autonomous, not to say automatic, behaviors. Traditions are associated with "frontiers" that are going to change. In the medical field, the idea of being treated away from home is becoming a reality. It should also change in scope. Conflicts and reluctance are numerous and clash with the description given in Rifkin's book. Therefore, thanks to the information exchanged through these new media, medical tourism is becoming a reality.

Another point we should consider has to do with copyright. The European Parliament, which has supported restrictive policies on copyright, has changed its views considerably. The concept of copyright is meaningful only in terms of the development of antitrust laws. The European Parliament has not made a global reflection on the policies that should be adopted in this domain. There is no need to be on the defensive. Approaches aimed at protecting the elements of medical knowledge and medical practices are establishing themselves, whereas others stick to the notion of the non-patentability of life.

The Internet is a new tool and its impact gives rise to arguments, as the telephone did before it. There are positive and negative aspects.

Positive effects	
The cultural dream becomes a reality	For example, knowledge and access to health-related culture
Negative effects	
Economically speaking, corporations are necessary	The system does not want people to possess medical equipment, obtain free medications, and lead their private lives

Table 7.1. *Positive and negative effects of the introduction of technologies*

We need policies in terms of technology since if we give free rein to the powerful, they will do what they want and not what people need. Certain societies literally steal culture to sell it again at incredibly low prices. Private life is being violated by cookies and, more recently, "web dogs", which are even harder to detect. The priority consists of finding out people's profiles. Microsoft is a monopoly that taxes its products. We buy usage rights rather than products. Even if unwillingly, we forcefully set up microtaxes that violate the normal law of the economy. Moreover, we do not punish those societies that manufacture bad products and we tax their products. Everyone accepts this since the cost of this mini-tax is much lower than the cost of the procedure.

The solution is political but we should also remember to analyze the technical means necessary to implement it.

7.2. The great "ICT and health" revolution

Revolutions take place when we:

– possess new forms of communication;

– know how to manipulate nature;

– change our modus operandi.

Man has been growing within the tautological dimension represented by the concepts of freedom and autonomy for a long time. Ownership is also an element of freedom obtained later on. Currently, freedom not only consists of autonomy and ownership, it also involves being one of the links of the network chain. Being free nowadays means being part of a network.

The way we manage nature has changed. The Internet and computers will enable us to control our genes. But who will be in charge of the genetic knowledge?

The advent of industry created left- and right-wing parties and, with them, a political New World associated with mass production. This access age modifies mass work and consequently this political division. For humans, marketable competences will never be the same.

Currently, we are at the beginning of a new era: the industrialization of medicine.

7.2.1. *Health and the Internet for everybody...*

We would be tempted to write: same difference! Europe would answer affirmatively to the question of "the Internet for everybody"; the problem lies in finding out what this involves and means. Internet access should be provided in public spaces, from libraries to healthcare establishments.

Telecommunications societies must have obligations in terms of public service. We should not allow the creation of a digital divide between those who are part of the system. The Internet must be accessible in free formats, which we should not be compelled to buy from companies. We have to pay attention to the disappearance of those companies that in turn, for this very reason, entail a loss of knowledge, since their dissolution can be somewhat compared with the burning of libraries.

The question would be: what does this mean in relation to the Internet? Let us consider the example of virtual reality: "if your children spend several hours immersed in this virtual reality, it is real". It is part of a specific form of reality that corresponds to its game dimension. This distinction between real and virtual is quite significant and will allow us to teach medicine differently. Therefore, surgeons will practice with simulation tools. This is something that, for the benefit of the elderly, could be expressed as "not on a patient for the first time".

7.2.2. *Which political developments should be implemented in the health industry?*

Political integration is necessary but there are also needs; we should also emphasize contentious issues and clashes. Society is often aware that something is not right before politicians, and often bureaucrats. We cannot do without political structures. On the other hand, the claim that politicians arrive late in their own field is untrue.

Back in 1944, Karl Polyani found out that if liberal economy works during periods of major change, it is thanks to social relationships.

Later on, Paul Seabright [SEA 10] proved this theory in his book *The Company of Strangers: A Natural History of Economic Life*. The other idea put forward consists of a development of the application of the notion of usefulness. This concept allows us to apply notions like signal, imitation, ostentation and status consumption. In terms of health, these notions are not very effective.

Perriault proposes an analysis of the logic of usage in relation to five parameters: Maurice Godelier's representation [GOD 97], Bourdieu's social norm of usage, the niche of usage, the structure of a project and the mark left by technology [PER 08, p. xv]. Michel Gensollen's analysis [GEN 99], which is more recent, is limited to three levels. It includes the recognition mechanism of the characteristics of experience goods. The second dimension of his analysis pertains to the creation of users and the learning process, a phenomenon that takes place thanks to the sharing of expertise and communities of practice, which constitutes the third level of his analysis. The creation of users is crucial in the health industry, where it is called patient education.

The social creation of tastes can be of two kinds: mimetic, as proposed by René Girard[1] – "the Other tells me what I like" – or cultural – "I like the support goods of my relationships with others". However, the trivialization of these products was quickly underway. Therefore, they have become just like any other goods product when they require a learning process and a familiarization, which decreases more and more throughout the development

1 The mimetic approach was founded by René Girard [GIR 61]. The version we consulted refers to Editions Hachette, coll. Pluriel, 2003.

of the products. This evolution is societal: if formerly a diabetic would hide to measure his dose, now young diabetics take out their pen dispensers in front of everybody. It is not only a banalization of the disease, but also a kind of social evolution that leads to forms of acceptance.

At this stage, we know it is important to develop the idea – presented in Jacques Perriault's work – that goes beyond the standard usage. This author defines the logic of usage. This analysis makes us think that Perriault and Gensollen's approaches are pretty much comparable. Consequently, we will use this definition of "usage":

DEFINITION 7.1.– *"The logic of usage consists of how an individual chooses a tool and employs it to carry out a project" [PER 08, p. xiv].*

This definition substantially takes up an analysis that the author will develop throughout his work:

> "Usage requires a double decision: buying the device and making use of it. In all the cases we have considered, we tried to pick out three elements involved in the decision and process of utilization. The first is the project. It consists of the anticipation of what we will do with the device. It can be more or less clear and acknowledged and it will often change when the object is used. The second is the device itself. The third is the function we attribute to it... For the user, the purpose of the device lies not so much in its functioning as in its use in relation to a service that has nothing to do with technology" [PER 08, p. 205].

In the next part of this book, we will dedicate ourselves to studying anticipated uses and their consequences. We will also wonder about the role played by these notions in the health environment, where rules and restrictions limit the users' choices.

7.3. Assessment tools

Experts have created certain tools to assess effectiveness in terms of access, usage and therefore performance. In the following section we will show some significant results to the reader.

7.3.1. *Population access*

If what is good for an individual x is also good for an individual y and, consequently, for a whole population P, then we talk about optimal population choice Pc. Let us consider now an individual choice Ic for a given patient i. We can calculate a ratio Ic/Pc. If this ratio is close to 1, then the individual choice is always equal to the population choice and constitutes the most favorable condition. On the contrary, this ratio gets closer to zero when the individual choices are very different from one another.

Luc Perino explains in his book that surgery is superior to all other branches of medicine since the ratio Ic/Pc is quite often close to 1. He considers two examples: peritonitis and the section of a part of the femoral artery. If we consider for instance a case of intestinal occlusion, one individual choice corresponds to two population choices, which yields an Ic/Pc ratio of 1/2. The choice can, therefore, be questioned [PER 03, p. 70].

7.3.2. *Relative risk*

This notion of relative risk (RR) determines the individuals that will be treated.

Relative risk is the ratio of future patients that present a risk factor to the total number of patients suffering from a certain disease.

Therefore, if the risk of the disease is double, RR will be equal to 2. This notion is very significant in that it will define public health policies like the battle against well-known risk factors. Thus, these public policies run the risk of promoting industrialization, often in the medical diagnosis industry. Currently, performances are more analyses of the impact of medications on these factors than considerations on lifestyle.

7.3.3. *POSSUM*

POSSUM (*Physiological and Operative Severity Score for the enUmeration of Mortality and morbidity*) was proposed as a way of knowing the risk of mortality and morbidity associated with surgical procedures on the digestive system. It is a form of risk assessment that is becoming more and more widespread.

7.3.4. *Conclusion*

As in a sort of Big Brother reality, citizens are transferring health-related information more and more; some manage it, but its owners remain unknown. In the best case scenario, when we control the past, we master the future. We have to think that this will also be the case for the health industry.

8

Mondialisation, the Maker of Biopower

The notion of biopower is born from the impact of *Mondialisation*. This happens because patients and healthcare professionals are becoming objects within the shape-shifting and global system of forces that constitutes biopower.

8.1. Mondialisation versus globalization (definition)

According to Marc Augé, "The word *Mondialisation* refers to two kinds of realities: on the one hand, what we call globalization, which corresponds to a worldwide extension of the so-called liberal market and of ICT networks, and on the other what we may define as planetary conscience or planeterization, which in turn is composed of two elements".

8.1.1. *Two kinds of conscience*

The topic we are discussing here, i.e. health, is deeply involved in these technological networks, and it is even their most tangible depiction.

In relation to the first meaning given, the diffusion of medical technologies around the globe relies on the same principle.

In this case, planetary conscience is a two-fold notion. The first connotation may be qualified as ecological:

> "Every day we are aware of living on the same fragile and threatened planet, infinitely small in an infinitely large universe".

This ecological conscience barely affects the health industry, apart maybe from its indirect effect on pollution-related diseases.

The second form of conscience is social:

> "We are also aware of the ever-growing gap between the richest and the poorest, and of the parallel gap between knowledge and ignorance".

What should we think of this planetary ecological or social awareness that is never put into action? In the health industry, this gap is inexorably becoming wider.

As a result:

> "The term 'globalization' refers to the existence of a global liberal market – or deemed such – and of a worldwide technological network, to which, however, a large number of people have no access yet. The global world is then a network-centric world, a system defined by spatial and also economic, technological and political parameters" [AUG 08, p. 41].

We will add psychological parameters to this list. The ancestral fear of those diseases that make the body decay from the inside because of external pathogens has two consequences: the psychology of disgust as felt for these pathogens leads to overdiagnosing, overdosing and overtreating.

8.1.2. *Alienation*

It would be easy to claim that the need for connection leads to a contemporary form of alienation, which is the mark of this new society. On the other hand, we will use several concepts that derive from these debates, which account for the following elements:

– The first consequence is that alienation becomes meaningless whenever a control-centric kind of culture makes us agree to be under the heel of a "Big Brother" or "Biopower". This is a theme that we have accepted together with Ronell.

– The second is that alienation becomes meaningless whenever permanent connection is indispensable and inexorable, which is the case for the technological systems that we can access on our mobiles, like social networks.

8.1.3. *Alienation according to Marxists*

We cannot ignore the concept of alienation without justifying its rejection. One of the first causes of this rejection is twofold. Often alienation is a sort of bar-room-politics concept. However, this notion of alienation as put forward by authors such as Marx or Hegel was re-formulated from their juvenilia all the way to their crucial works, without any sort of detectable stability in the matter.

The second reason for this rejection lies in the alienation/work association. As François Perroux recalls: "K. Marx, Hegel's heir, fights against his legacy. He removes the concept of first alienation from the political domain and that of disalienation from the world of ideas. He finds the first and fundamental form of alienation in the economic dimension. It derives from the relationship between capital and labor" [PER 70, p. 8]. It is therefore clear that the capital/labor relationship is quite different from interpersonal associations or the connections created between an individual and a machine by means of a mobile that links them together. Nonetheless, our views throughout the text will be informed by Marxism. According to Perroux, "Marx's terminology itself hints at different kinds of alienation. Production becomes alien to man. Labor and its driving force turn into objects. Social relationships become reified and objectified". This last point will become the focus of our analysis and justifies our ontological and axiological approach. How are mobile phones, as objects, reified? How does the concrete telephone relationship between two people become a "thing" that can be analyzed as such? The literature on this topic will show us that this reification can be carried out in different ways: for example, through images according to McLuhan and through voice according to Ferraris[1].

The basic form of Marxist alienation can be found within infrastructures, but there is a secondary form of alienation located within superstructures.

1 A significant part of this theory from here onwards will be considerably influenced by Maurizio Ferrari's reading, adopted in his work [FER 06].

Consequently, complete disalienation would require the destruction of both these types of alienation.

8.1.4. *What role does alienation play?*

In the previous sections, we rejected the notion that the alienation deriving from addiction is specific to a technology, such as video games, or constitutes one of its features. On the other hand, the notion of boredom leads us back to this topic.

Boredom always involves the awareness of being trapped in a given situation or, quite simply, globality. We could qualify this imprisonment as alienation. Alienation cannot be an ideal, as that would raise the question of knowing what men are alienated from. It requires us to spell out another element: the alienating object, namely, the element that has been lost. The cause of alienation does not lie in mobile phones or video games, but rather in the situation we find ourselves in.

According to Bertrand Vergely, alienation is characteristic of a condition of self-estrangement. This situation allows us to understand the human condition that consists of no longer being what we are without having lost our being [VER 03, p. 76].

According to Svendsen, boredom has a dehumanizing effect, which is a type of alienation. It deprives human life of the meaning that actually makes it what it is [SVE 03, p. 284]. Our existence is constituted by our being-in-the-world or, in other terms, a polarity between subject and object, men and what surrounds them. Svendsen also claims that, when bored, we feel "the inexistence of reality or the reality of inexistence".

These two authors lead us to wonder about the alienating role played by the aforementioned situations in the health industry. Certain members of the medical profession confine this alienation to the mere constraint represented by the regular intake of medication.

8.1.5. *François Perroux's critique*

In his work, François Perroux [PER 70] introduces the term objectification to contest this approach which reduces reality to objects. In the unindustrial system of medicine, which is going to disappear, objectification is a key notion. This process lies beneath the concept of evidence-based medicine (EBM) or biomarkers.

> "Some systematic reflections mistake objectification for alienation. Every objectification of the subject is regarded as a reification and concretization which, as a consequence, alienates him. Now, every subsistent, existing, and coherent subject consists of objectification; if every objectification were a kind of reification, every discernible subject would vanish and the fact of analyzing it would throw us into the dimensions of things" [PER 70, p. 19].

François Perroux claims that "in every direction, the individual is desubjectified by a system of institutions (game-rule institutions and body institution). The irreducibility of self-knowledge and the originality of one's spontaneity can only be grasped in relation to networks of social specification and determination" [PER 70, p. 22]. Clinical trials sounded the knell for individual medicine. Faced with the arrogance of some mandarins in the know, clinical trials are becoming one of the mainstays of budding industrialization.

This quotation leads us to the core of our approach. Perroux himself provides us, in the following pages, with some method guidelines:

> "At any given moment, each subject is objectified and socialized: 1) by his social horizon, i.e. the number of variables he takes into account to set up his existential plan or abandon a given plan of action; 2) by the scope of his possibilities, namely the number of alternative situations that he compares to his previous situation by deeming them achievable; 3) by his sphere of action or influence, namely the number and kind of variables that he modifies by performing conscious and deliberate actions. From time to time, these three spheres happen to be interrelated" [PER 70, p. 24].

Due to this fact, new choices become available. No doctor knows what is best for a given individual, but he knows the result of clinical trials carried out on groups of symptoms (at least theoretically). Moreover, no theory will be able to verify a clinical result for the whole of the population, since each individual is different.

This point of view seems too mechanical to us even in relation to cybernetics, even if later on we will be led to analyze the scope of possibilities. Randomized clinical trials are carried out in order to eliminate all kinds of subjective factors and even the subjectivity of the patient itself, which is often called placebo effect.

8.1.6. *The recent rejection of alienation-improvement duality in the health industry*

Every journal belonging to the literature of the early 1980s is characterized by the multiplicity of texts that it proposes in relation to this field. These documents constitute a critical response to the myth of technological progress. The most recent type of this debate consists of the search for a point of reference between alienation and the well-being created by its objects. A recent form of rejection of this kind of debate can be found in authors like Caron and Caronia who, from the very introduction of their work, announce that: "As communication instruments become part of our daily routine, they free us from most of the spatial and temporal constraints that govern our lives. Significantly, overcoming such constraints has ramifications beyond the ability to effectively manage the multiple, simultaneous tasks that characterize contemporary life" [CAR 05, introduction, p. 6]. A few years before, François Perroux was already mentioning this debate: "What is worse, devices seem to make use of men. The alienated man keeps describing his anxiety in revelatory terms: he defines himself as a cog of a large machine, he is caught in a mesh, he is crushed by the machine (a hellish machine in which ignorance and fate work against men, as was the case in Jean Cocteau's famous work); he is laminated like metal, snatched like a drive belt, crushed by a weight like in an elevator crash" [PER 70, p. 32]. Clinical trials are based on the principle of biostatistics in order to eliminate any kind of alienation limited to a few individuals. The best choice for a person consists of a group choice. Thus, we get to the analogy between labor and the car industry assembly line. The principle of EBM, a statistical form of medicine, is "let us automatically treat all those patients with the same data", which constitutes the basis for a creeping and hidden form of industrialization. EBM is paradoxical in that it attempts to treat patients when they actually aspire to a customized kind of medicine. This demand is even more accentuated by our *santacteur*.

This debate is recently being revived, even if no longer in terms of alienation versus improvement. It has more to do with the impact of these technologies on spatial and temporal constraints, and with the role played by these tools in present-day life.

As for the first point, Caron and Caronia affirm a few lines further on that: "These are examples of the many ways that emerging communication technologies impart meaning to daily life. Once they become integrated into our routines, they reformulate possible meanings: places, actors, reciprocal relations, and the typical events that comprise them are thus amenable to new modes of accomplishment and interpretations" [CAR 05, p. 7].

At first, this creates a research approach: places, actors, their reciprocal relations and typical events. In a second phase, it brings back into focus certain authors that had become obsolete, such as Maslow and his levels of achievement.

Variables	Comments
Places	Determining the places where treatment is available
Actors	This notion recalls the concept of digital identity that we will deal with throughout this book
Relations	It is at once a philosophical and pragmatic kind of basis: why do humans engage in relationships?
Events	Which events generate usage? What is the role played by present reality?
Accomplishment	What is the level of accomplishment?

Table 8.1. *Approaches for the analysis of communication according to Caro and Caronia*

8.2. Are there temporal paradoxes?

Marc Augé describes three temporal paradoxes. He defines time as a limit, i.e. someone's lifespan, but also in terms of events. We will use these notions, as we continue to analyze our relationship to time.

8.2.1. *Marc Augé's three temporal paradoxes*

Marc Augé's three temporal paradoxes are defined as follows:

> "The first temporal paradox is inherent to the individual's awareness of existing within time from his birth and well after his death. This realization in terms of finiteness and infinity applies as much to the individual as it does to society".
>
> "The second temporal paradox is in polar opposition with the first: it pertains to mortal men's difficulties, i.e. individuals who rely on time and the notions of beginning and end, in conceiving the world without assigning a birth and an end date to it".

This notion is represented in the health industry by the care pathway, among others things.

> The third temporal paradox has to do with its content or, in other terms, history. It is the paradox of events, always expected and always dreaded. On one hand, it is events that make the passing of time perceptible. They are even used to assign a specific date to it and to organize it in a way not related to the mere cycle of the seasons. On the other hand, events involve the risk of rupture, of an irreversible break with the past, and of an irreparable intrusion of innovation in its most dangerous forms" [AUG 08, p. 7].

Among these events, let us highlight the so-called life incidents!

Out of these three paradoxes, the one that has to do with events seems the most interesting.

8.2.2. *Time versus space*

One of the points on which Marc Augé and Giorgio Agamben agree is the impossibility of separating time from space. Marc Augé writes: "The fact remains that it would be pointless to separate a reflection on time from space-related considerations. All the symbolic systems that we can observe everywhere demonstrate a form of solidarity, always intuitively perceived, between these two *a priori* forms of our sensibility" [AUG 08, p. 11].

In this situation, where the concept of a "global village" remains the dream of some experts, there are surprising statistics that show this conjunction between time and space, as it is described by these two authors. In spite of the possibility of extending new social networks to a global level, the research carried out by Kissmetrics shows that, for most people, friends are physically close and that we are getting into temporal kinds of habits more and more often in our relations with them. This notion plays a significant part in the possible reflection on the role of careers in terms of relationship with healthcare.

8.2.3. *Biopower, diaspora, power and territories, and disappearing borders*

We are aware that technological products give rise to "religions" and consequently beliefs. The same goes for the health industry. Not only do these beliefs constitute the basis for cultural forms, they also create new types of power. We will mainly tackle the issue of biopower, which Mattelart prefers to call the "surveillance society[2]". This notion of surveillance is one of the features of the new society based on mobile phones [MAT 08].

2 The exact title of his work is *The Globalization of Surveillance* [MAT 10].

9

Belief, Myth and Biopower

Biopower takes shape thanks to, if not exclusively through, beliefs and myths. It may appear as a form of objective opposition to variables or as the will to universalize rarely used populational medical choices. Biopower can also manifest itself as *a priori* stances, often validated by the principle of precaution, which do nothing more than sterilize any form of scientific progress.

9.1. The problem of belief

Several authors focus on this notion of belief as a source of culture and identity. We hope we will not turn it into the pivot around which all of our analysis revolves, even if the role it plays is often regarded as central. Here, we focus on current medicine, which claims to scientifically reject all kinds of beliefs in its domain.

9.1.1. *The current role played by beliefs in the health industry*

"The concept of belief is far from being a blind acceptance of something unexpressible or supernatural. It involves accepting principles that structure daily life. If we follow this logic, then beliefs have to do with what I defined as "the social sense". By sense I mean the relation between one and another such as it is conceivable, conceived, represented, and eventually instituted" [AUG 08, p. 87].

Evidence-based medicine (EBM) exists in a sort of fight against beliefs and, in particular, the omnipotence of the bureaucrats of the past. We should point out that this dimension is not that categorical, and clinical trials can be used by beliefs themselves. Jauréguiberry and Proulx admit that "strategic calculations can be made to serve a cause, a commitment, or a belief" [JAU 12].

Any belief defines the transgression of desires, pleasures and taboos. Hence, it also defines "powers". We will point out later the significance of the biopower generated by the identity created by using tools. We will also consider the potential despotism of this power.

The concept of belief is tricky to use, as it is also in line with the dynamics of Marxist discourse. According to Marx, every kind of society is made of an infrastructure and a superstructure. The former corresponds to the mode of production itself and works according to the rationality and effectiveness described previously. It consists of production means, workers and capital or, in our case, healthcare professionals and establishments. The superstructure is determined by this infrastructure, which is the actual motor of society. It consists mainly of beliefs, cultural forms and the system of laws. According to Marx, each type of society is determined by its beliefs and culture. These two elements interact and create what we often call medical ethics.

9.1.2. *Biopower*[1]

The question we will tackle in this chapter pertains to the secret or hidden role that medicine and its actors may play. Medical biopower has become a tool that society as a whole uses to exert its power on the individual. This type of power cropped up during the analysis of the relationships between ethics and technology. It is called biopower. It is linked to a notion that we will name "trace"[2]. This concept of biopower is in line with that of an inscription or mark, which becomes the indicator of an event. This is when

1 This section is a reflection that took shape during the research seminar called *Etos Technologies, contrôle, démocratie*, organized by the Institut National des Télécommunications, the École des Hautes Études en Sciences Sociales and New York University. It only hints at the debates that informed that seminar.

2 The famous detailed invoices that we have already talked about are examples of the marks left by man by means of technology.

the concepts of medical diagnosis and surveillance society come into play and become significant for this approach.

The target of traceability is the achievement of security, but this notion also raises problems of identity straight away in the medical field. It aims to better qualify the actions performed and the products (protocols, medications and devices) launched on the health "market". It allows us to exert more control over the risks involved and contributes to their decrease. Traceability becomes necessary if we want to find the source of the problems. It is increasingly required by norms and certifications. In this case, it consists of guarding against difficulties in terms of responsibility. Due to the multiple levels introduced by supply chains, this action significantly increases data exchange. We have to be careful at this point. We should not confuse traceability, which is a strict and objective process, with medical reports, which constitute a literary and subjective representation of the work carried out on a patient.

9.1.3. *The secret power of mediation*

Supposing that the media have a secret sort of power implies refuting a theory of communication called "agenda setting", according to which the media cannot modify or fashion behaviors. This theory limits their usefulness to the sharing of preoccupation. Therefore, they could only sort out social priorities [MCC 72].

At the beginning of this work, we approached the concept of socialization by basing our analysis on the claim that the media are involved in it. An intermediate position could consist in stating that the media may require a certain form of social reality. The question then boils down to understanding whether it is a form of power that affects human reality or the organic dimension and how new medical technologies contribute to it.

9.1.4. *What is biopower?*

The notion of biopower leads us back to the fact that man unwillingly leaves some traces behind him. These can be of different kinds. We can find them on his computer, credit card, health insurance or transport card (Navigo card for Paris). We can also find them, thanks to biometry, when he crosses a border or thanks to the SIM card in his mobile phone, which also records his blood sugar levels by means of a blood testing device. The multiplication of the technological solutions used in the health industry certainly constitutes a new source of traces and, consequently, a basis for the expression of a type of biopower.

The abstract of an article by Katia Genel [GEN 04] illustrates this perspective quite aptly:

> "According to Foucault, a transformation in the exercise of power comes to light beginning with the 18th century, as life itself becomes an object of concern for power. 'Biopower' is the term he uses to describe the new mechanisms and tactics of power focused on life (that is to say, individual bodies and populations), distinguishing such mechanisms from those that exert their influence within the legal and political sphere of sovereign power. In *Homo Sacer*, Agamben takes up Foucault's analysis and re-establishes it on the very grounds that the latter had wanted to break from: the field of sovereignty. Agamben argues that sovereign power is not linked to the capacity to bear rights, but is covertly linked to a 'bare life', exposed to the violence and the decision of sovereign power".

Foucault's theory is a double-edged sword. First of all, biopower can only be understood in terms of modes of opposition. It is a form of power relation. These relations of power are above all tools at the service of knowledge, and once these two elements – power relation and tools of knowledge – are present, they can become a reality.

According to Pierre-Antoine Chardel [CHA 09], the reflection on digital marks and traces, i.e. the evidence of this biopower, is more in line with Hobbes' approach than Rousseau's. In 1651, Hobbes explains in his *Leviathan* that the purpose of the state is to ensure social peace. The reason behind this is simple: "man is a wolf to man". The most important notion is the value of security. Rousseau objected that men also live in solitary confinement, under a state of meticulous surveillance. In the first book of his *Social Contract*, his intention is to transcend the protection of the goods of each member and to claim that security is no longer the sole priority. It has to be coupled with freedom. Therefore, he claims the existence of a political ideal alongside a moral one.

Armand Matterlart states this in his book about the surveillance society. The techniques used to trace individuals detect their behavior and consumption. This results in marketing practices of this unofficial information or in executives making use of it. This is what leads Armand Matterlart to state that a global surveillance society is being set up [MAT 08]. This matches the advent of authoritarianism established by this surveillance society. According to the theories proposed by Marx or the Frankfurt School, this despotism takes shape within a society through the renunciation of positive reason in favor of instrumental rationality by means of technical tools. In this kind of society, individuals do not try to find out if something is useful or allows them to reach their goal. This is supposedly the reason why people install on their mobile phones those applications that monitor them: because they are useful and allow them to achieve their goals. The whole matter revolves around the duality of this operation or the mediation between the service provided and the risk of being monitored. The proliferation of well-being apps, based on sensor wristbands of all sorts, is one of the results.

Currently, one of the trendy terms is "big data". "Data mining" allows us to process all of these data, make predictions and describe patterns. The hard part lies in what the following example illustrates. Someone's application for a loan is rejected by a bank that has done some "data mining". This bank reckons that this person is affected by cancer, who in reality has actually done some research on the disease of a friend who, because of his hand tumor, was not able to type on a keyboard.

The concept of power no longer designates a system of rules, but the set of methods used by the normalizing power. This claim gives rise and meaning to the study of traces. The analysis of the concept of biopower also becomes significant.

9.2. *Critique of the notion of biopower*

Katia Genel intends to "bring into relief the extent to which Agamben shifts the meaning and content of Foucault's notion of biopower [...] and this notion of power when applied to sovereign power, in order to assess its relevance and fruitfulness".

Castoriadis' analysis [CAS 75] is quite different. He opposed Foucault's questions about the resistance put up by certain singularities to control techniques. In 1963, he starts advocating the resumption of the Revolution. He introduces a separation between dominator and dominated. He sees society as built entirely on the distinction between leaders and subordinates. Therefore, he regards biopower as the dominance of an enormous bureaucratic structure which shows a lack of popular sovereignty.

In 1976, Foucault introduces the notion of "will to knowledge". He supports this theory in relation to complex and unmastered power relations. He denounces the illusion held in place by the power the high ranks exert on their subordinates and puts forward in its stead a system of balance of power in which no one can force himself upon others. According to him, our condition is characterized by the omnipresence of power: power is everywhere. Therefore, every imposition of power necessitates a phenomenon of opposition. In this system of balance of power, it would therefore be naïve to abolish the power of the state.

The different notions of power put forward by Castoriadis and Foucault can give rise to productive considerations. Castoriadis, unlike Foucault, who supposes the existence of a state, allows us to define society as a whole as the source of this power. A first expression of this biopower can be seen in the bandwagon effect we described at the beginning of this book. In other terms, the biopower of society as an entity requires the individual to own a mobile phone.

9.2.1. *Biopower and psyche*

Nicolas Poirier [POI 06] evokes the wicked fantasy of voluntary servitude described by François Ewald to position his line of thought more precisely in relation to Castoriadis' philosophy. Therefore, those who are dominated have only one solution: opposing power. If Foucault separated the two opposites of living and political mankind and saw a complex system contained within the two ends of the spectrum, Castoriadis regards humankind as a monstrosity and social history as the usherer of modernity. Therefore, according to Castoriadis, it is social institutions which create human beings, which is a theory that Nicolas Poirier seems to take up. If we follow this logic, the kind of society that we previously described, which forces itself upon the individual through the use of mobile phones, becomes an institution.

Nicolas Poirier introduces then the notion of psyche. The hypothesis at the basis of this concept is that the individual is the product of a process of socialization. Social issues are a game we can play only once. On the contrary, psyche is linked to an "original fantasy", the monad of libido and the omnipotence of the womb. The newborn is the carrier of an unacceptable imaginary power. Hence, the socialization of the psyche is necessary for the creation of a social individual. Therefore, the integration of psyche into a social dimension creates the individual. Standardization takes place through education and the use of training tools, which are also ways in which the psyche can access meaning. On the other hand, according to Nicolas Poirier, Agamben's "bare life" is unrealistic, not to say impossible. This operation of socialization of the psyche, which Castoriadis calls "sublimation", is seen in a positive light. According to him, no human decides to associate himself with others. Any being or any individual is situated in a given framework. Within this compulsory wholeness, individual acts are possible. According to Nicolas Poirier, biopower technicians all make the same mistake. Every new society that makes use of technologies like mobile phones reconfigures biological life to make it its own. In simplified terms, what justifies fighting is the presence of the "lens of singularities" as opposed to "health standardization". The matter boils down to knowing against whom or what we fight.

A philosopher may be criticized for his slightly exaggerated view when he claims that what mainly distinguishes Foucault from Castoriadis is the trust in institutions. "Every man is part of institutions". According to Castoriadis, "institutions are linked to the praxis of a "creative institution"".

Thus, Nicolas Poirier considers the existence of a positive version of control. According to him, the answer lies in the question of men's modes of opposition. This is why Foucault and Agamben use different approaches in order to adopt a point of view which is in polar opposition to Marxism.

9.2.2. *Biopower according to Foucault*

According to Foucault, power is the sovereign's right to rule. Biopower controls vital processes. It is a health approach suitable for a process of standardization: "biopower keeps us alive". Therefore, Foucault falls in line with Lukacs [LUK 60] with his notion of class consciousness, although he wavers from one end of the spectrum to the other.

At first, Foucault focuses on the individuals that power – a disciplinary kind of power – tries to make obedient and productive. He creates an "anatomo-politics of the human body". Thus, the body must be trained so that it can incorporate the norms of capitalist production. It is an individual-centric body discipline. Foucault's disciplinary power was certainly an aspect of the history of telecommunications. Pagers allowed groups to order their militants to do certain things. However, the impact of this disciplinary power has remained weak. We could also mention ankle bracelets, namely, devices attached to someone's body which use a mobile phone-like technology that allows us to find out the position of its wearer at any given time. At the other end of the spectrum, man belongs to a living species that has to be controlled biologically. This is how mass birth and health control and no-smoking campaigns are justified.

Foucault breaks away from traditional political order altogether. Judicial power was formerly in the hands of the monarch, who decreed what was allowed and what was prohibited. He also had the power to take someone's life. Current legal norms allow us to exert control over life and define sane behavior, even if scientific progress proves the exact contrary. Therefore, it is in our best interest to remain healthy. Similarly, offenders become sick people and sanctions must be replaced with appropriate treatments.

Nicolas Poirier regards democracy as an explicit and permanent self-institution of society. Christian Ruby [RUB 07] claims that Nicolas Poirier offers a line of thought that "structures its coherence around a theory of the imaginary characterized by: constant attentiveness in relation to political issues, avoiding conceiving institutions as social things, etc.". Institutions must ensure freedom and economic life (social control), and they must enable the ruler to increase his power.

Thus, Nicolas Poirier's point of view is pessimistic. Monitoring our bodies fits into the attempt to widen the scope of power so that the individual himself can be brought under control. Biopower leads naturally to a totalitarian dimension, where these tools try to make power even more effective.

9.2.3. *The denunciation dogma of moralism*

Nicolas Poirier condemns first of all the "denunciation dogma of health moralism", which is defined by constraints: "to stop smoking", "to stop drinking", etc. His reply is the following: "Doctors have too much power". Therefore, man becomes a slave to a technique, namely medicine, which has turned against him, hence the importance of the condemnation of this new health order. What does this entail for mobile phones? As we have already pointed out, ordering people to stop making calls is pointless.

Poirier proposes to leave behind health policies, as our Western societies are becoming control-centric. According to him, biopower stands out more for its absence than for its omnipresence and can no longer ensure living conditions to the poorest classes, as not everybody has the same opportunities in terms of healthcare. According to him, there is no equality in the face of biopower. It is no longer a form of control as a necessity that should be accessible by anybody. However, the vast majority of people have no access to control. It would therefore be wrong to consider the capitalist system as a will of biopower. Poirier thinks it is only the daily activity of the exploiting classes that led us to provide an answer to these needs. If we follow this line of thought, we still have to establish which classes are "exploiting".

Standardizing biopower no longer exists. If it were functional, it would be entirely associated with the capitalist state. However, this is not the case, as Nicolas Poirier explains, when considering the examples of social security and pension schemes. Once again, the theory put forward by Nicolas Poirier is undermined.

9.3. *Equality and tyrannical kinds of power*

John Rawls, in his Theory of Justice [RAW 05], subordinates equality to freedom, even if starting from a "basic income level". Therefore, freedom can be sacrificed in situations of underdevelopment. Amartya Sen's [SEN 73] line of thought is essentially the opposite. According to him, we have to accept the principles of inequality in order to relate to Castoriadis and observe the news forms of power.

Castoriadis thinks that the current situation is defined by oligarchies that dominate the economy. Are we talking about the traditional bureaucrat? There is undoubtedly more to it! As a consequence, according to some, man cannot be simultaneously dominant and dominated, which contradicts the complexity of power sustained by Foucault. Castoriadis is in line with a discourse of opposition resulting in a polarity that allows us to define friend and foe. Consequently, if power is disseminated, it is also favored. This is the reason why we have to establish who is dominant.

Should it be otherwise, conspiracy theories would become central. Nicolas Poirier sees only two solutions: either we have a total diffusion of power or we trigger opposition by pointing the finger at the enemy. It is at this stage that we get to the key point: the user has no enemy, only degrees of relation.

It is true that Foucault wrote these works in a period when two forms of power, in this case, financial and technological power, were less repressive than they are nowadays. The coincidence of these types of power and *mondialisation* means that the dominant-dominated problem arises more on a local basis, if within a complex geopolitical system. The concept of biopower was born in the 1970s when wild capitalism had not taken place yet. Consequently, we should integrate the build-up of financial and technical power into this analysis.

Currently, many philosophers, following Beck's example, are talking about a fake "Big Brother" pattern. Beck also considers a future society going through a process of *mondialisation*. According to him, cosmopolitanism may be the value of tomorrow, which is something that alters the relations of power. Thus, there are individual solutions that are becoming more significant than their collective counterparts. The dynamics change and we are consequently faced with risks and uncertainties.

On the other hand, Jean de Maillard[3] reckons that "the most paradoxical aspect of *mondialisation*, which brings together the space of exchange and communication on a nearly global level, consists – in sociological terms – in an indeterminate and uncontrollable dispersal of the forms of socialization".

3 Jean de Maillard is a judge, blogger and writer of essays that captures the attention of the media such as *Le Rapport censure: Critique non autorisée d'un monde déréglé* [MAI 04].

This results in scattered centers and the absence of a "Big Brother" which, when applied to our topic, becomes more pertinent notion.

9.3.1. *Biopower and institutions*

The change taking place in human institutions is a mainstay of current evolution. Foucault regards it as an anthropological mutation. It consists in the shift from the omnipotence specific to a group or caste to a form of omnipotence that anybody can gain access to. Thanks to our mobile phones, we feel vested with power when we access information, interact with someone or wish to know where we are. The same will happen in relation to future health applications. According to Foucault, it is law that makes an individual. So, he necessarily refers to institutions whereas, for Habermas, law establishes life. This distinction is fundamental to the current debates we may have about the end of life.

What kind of education could favor this or that imaginary institution or that imaginary life by means of medical technologies? This could be the question, which quickly clashes with the inability to "step down". This is why society necessitates the creation of "momentary" institutional forms.

The significance of Castoriadis' works lies in how he shows that no coherent opposition can exist without being involved with institutions. Therefore, he is opposed to Antonio Negri's notion of "multitude", "which is not an institution" and corresponds more to the multiplicity of uses that currently manifest themselves as a multiplicity of applications. We still have to define its uses and applications. According to Antonio Negri and Michael Hardt [NEG 05], we are living in an "Age of Empire", which corresponds to postmodernity. The power of the Empire is stifling us, and Antonio Negri and Michael Hardt[4] are making an appeal to the "Multitude".

9.3.2. *Biopower and imagination*

Castoriadis's line of thought conceives a situation of mastery over nature, where medicine would eventually fight all kinds of diseases. He founds his reasoning on Foucault's notion of rational mastery. Whereas Foucault

4 This is the interview we refer to: http://multitudes.samizdat.net/Interview-de-Toni-Negri-et-Michael.

concurs with the notion of rationality, Castoriadis assumes that mastery is pseudo-rational. This ideology of mastery over nature, allowed in our modern society, also seems to be a form of global mastery, as locating systems enables us to control the territory. We mentioned this rationality at the beginning of this text, and it remains therefore one of the mainstays of the use of devices.

Let us go back to the diagnosis of our period, which Castoriadis uses as the basis for his imaginary approach. His is a fluid world made of states and an elusive American empire. From 1961 onwards, Castoriadis elaborates the fundamental concept of "Social Imaginary Significations" (SIS). He describes them as follows: "Significations are that by means of which and on the basis of which individuals are formed as social individuals, capable of participating in social doing and representing/saying, capable of representing, acting, thinking in a compatible, coherent, convergent manner, even if this be in the form of conflict[5]" [CAS 75, p. 528]. Therefore, the fundamental question would be: "What do we do?". It is also accompanied by some additional questions: are our mobile phones involved in the creation of these SIS? Should we fight for nation states or welfare states? Should we start pockets of resistance everywhere?

The weakening of the power of the state in favor of the economic one is a reality. The loss of political power is accompanied by the increasingly significant role played by stockholders and finance. The private sphere carries more weight than the states responsible for health. This situation is becoming widespread. This is the case for Europe, which first establishes itself as a large market and only afterwards attempts to set itself up as a political system. This leads us to the consideration that we should strive to create a different Europe and, in particular, European health (which has yet to be built).

This is also the case for what we may regard as the "sticky situation" of Western ideology, which has destroyed the concept of family and now wants to take charge of the elderly. Gerontechnology and medical industrialization will not be able to sort everything out. Moreover, states struggle to provide healthcare access to low-income individuals. The French Universal Health Care (*Couverture Maladie Universelle*) or Brazil's Unified Health System (UHS) fall within this logic. On the other hand, we should point out the

5 This quote is drawn from his famous work *The Imaginary Institution of Society*.

difficulties faced by Americans in this process of generalization. Is it a lack of resources? Mobile phones are used to keep in touch with people affected by Alzheimer's and find out their geographical coordinates.

These examples raise once again the problem of the role played by the state as imaginary institution. Can we conceive a non-sovereign state that manages not to disappear? This kind of state would favor the advent of increasingly customized mobile applications. The first approach would be political. How could mobile phones contribute to the resolution of political problems? How could citizens use them to face the problems of modernity? Could we give back to these institutions their function without turning them into a repressive state? The second approach involves the transparency of administrative data, or "open data", created by a decree in 2011[6]. It is a matter of open access to administrative data[7]. How could transparency allow us to create new applications which are useful for the citizen? We point out at this stage that the discussion boils down to the political dimension. "Open Data is the foundation of a firmly embedded and larger concept, open government, which aims to involve citizens in decision-making and to allow them to follow processes of implementation"[8]. These are all questions we may ask, but none of them can yet be answered.

The critique of this approach is theological. The state is too close to us. Could there be a form of opposition? If so, then it would certainly be similar to a sort of faith. A faith is a moral authority that lays down principles and values. It passes them on to technologies through education. Computers and mobile phones can illustrate this point. A mobile phone user belongs to the Apple, RIM, or Android "faith". Inevitably, this will be the case for healthcare technologies.

---

6 As soon as it was published in the *Journal official* on 31 May 20011, Decree-Law no. 41 of 26 May 2011 regarding public access to administrative documents (open data) was modified by Decree-Law no. 54 of 11 June 2011 regarding the Administration's responsibilities in terms of refusal of access.

7 "Potentially all data possessed by administrative bodies over the course of their public service task. Ministry, public organism, territorial collectivity…it is a matter of geographical, statistical and epidemiological information as much as of transport timetables, catalogs, yearbooks, etc. We should point out that this data also pertains to their inherent activity or, in other words, their management or the markets they have interacted with." Drawn from: http://www.lesnoubasdici.com/2011/07/open-data-dix-questions-pour-tout-comprendre.

8 This was drawn from the blog http://www.regardscitoyens.org.

What ensues is a kind of human self-limitation. It is a strong idea which emerges from these new types of states, where man takes his place in the world with regard to a set of technologies such as cars or mobile phones.

Could we then make a suggestion? Reality educates all of us through what we call experience. However, no one can make a state out of experience. This is perhaps why we see the emergence of the terms liberal oligarchy or participatory democracy. They are accepted because no one is being rigorous, but their tools give them the potential to develop. Hospitalization is often regarded as a negative experience, whereas hydrotherapeutic cure has often positive sides.

9.3.3. *Belief and mutation*

Genetics uses the word mutation to describe the evolution of our genes. Popular beliefs attribute two meanings to this term.

Its connotation is generally very negative. Mutations are the material of bad news such as pathologies, but they are often also associated with disabilities. Therefore, mutations are linked with diseases.

Mutations constitute the basis for life, allowing men to adapt to their environment. They are the source of evolution and are responsible for the survival of the human species.

9.3.4. *Aging well, treated well*

Prevention is the most important factor in terms of aging well, and technology can make the job easier for us. Evidence of this can be found if we consider patients affected by Alzheimer's, who require a lot of attention. Currently, it has been proved that certain stimulation tools are at least as effective as some pharmacotherapies. These technologies will therefore make the life of carers, who show actual signs of burnout, easier. In this case, the carer often dies before the patients. Therefore, we should propose that companies and associations should help carers by means of technologies, which certain structures like the *Université des aidants* are currently doing.

"The problem lies in accepting that we grow old". The first type of ill-treatment is isolation. This is a major concern related to insufficient or lacking housing adaptations. Inadequately adapted houses constitute one of the elements that need to be taken into account in relation to the ill-treatment of people who are dependent and find themselves more and more isolated, and who are for instance unable to go out.

There is also ignorance about finding assistance. If we consider again the previous example, in France, less than 50,000 modifications per year are compatible, which is a very low figure in relation to the needs (source: Pact Arim).

Ill-treatment also depends on those who deal with the elderly or a dependent person. Alice Casagrande, who focused on this topic, associates it with different hypotheses and observations. According to this author, we should learn from what we know about ill-treatment to progress toward suitable treatment. These health-related jobs are first of all depreciated and feminized. This entails a sort of gamble in terms of training, hence the opening of the huge market of training in the field of suitable treatment. However, this neither solves vigilance-related issues nor boosts the long-awaited feeling of trust.

Out of several types of abuses, whether organizational or institutional, this is the one that gives rise to individual ill-treatment. This is the reason why the technical systems introduced by establishments can be a source of abuse validated by the institution. We should also recall that there are many careless people in an environment where the majority of workers try to do their job properly, as happens in all other environments.

9.4. *Conception or has man become yet another object?*

The hypothesis that prostheses are a response to boredom is in line with the process of banalization of men, who fit in this network as vulgar objects. This common discourse is part of a network-centric world that introduces lifecycles and makes man immortal on the basis that he leaves a trace. Michel Serres explains willingly how men at first would only leave a trace, thanks to their ejaculate. The development of writing systems increased the number and nature of traces. As of now, these marks live on after men have died, which is clearly a novelty. Being part of a network involves leaving a mark. Therefore, man is an object made up of a set of traces, one of which is his health condition.

9.4.1. *Toward the immortality of the new man*

Jacques Perriault tackles the topic of immortality in relation to the idea of stopping time. According to him, "communication devices are also associated to a temporal type of metaphysics that concerns death and immortality. We can also secure pseudo-immortality through effigies" [PER 08, p. 68]. We should elaborate on this point, as technological proliferation and the intricacy of functions lead several factors to bring about new types of immortality, especially the technological progress made by networks which allow us to keep more traces and the constant quest for forms of identities. This kind of man, constantly fixed, will leave some traces. His prosthetic arm will have the potential to be reused by another man or to accompany him to his grave, similarly to how Egyptians left domestic objects in the tombs of their close relatives.

Besides this kind of immortality created by marks, we should also mention the one brought about by the aforementioned mutations. Thus, some researchers dream that some of these genetic mutations will make men actually immortal.

9.4.2. *The technological progress of networks makes immortality possible*

Bernard Benhamou[9] asks the following question: there are 1.3 billion people connected to the Internet, but how many objects? We would rather focus on the 7 billion people who inhabit this world and the 5 billion people who own a mobile phone.

In order to analyze these phenomena, Bernard Benhamou starts from the principle that the Internet as a network must be analyzed in relation to three dimensions. According to him, three approaches should be used: a technical one,

9 This part is drawn from Bernard Benhamou's conference "*Les Entretiens du nouveau monde industriel*" held on 3–4 October 2008 at Centre Pompidou, Paris. It was organized by the Institut de Recherche et d'Innovation (IRI)/Centre Pompidou, the competitive cluster Cap Digital and the ENSCI-Les Ateliers (Ecole Nationale Supérieure de Création Industrielle).

which regards the Internet as a set of interconnected machines; a human one, which sees it as constituted by groups of people; and an informational one, which focuses on its content. According to Bernard Benhamou, organizing and implementing these resources constitute political choices.

He proposes then a reflection on technology. The principle itself that regulates Facebook consists in the obliteration of privacy and a form of exposure traveling outward in concentric circles. This same process also characterizes Google+. More than 3 million people have bought "NikeiD", a communication system that uses a chip to link the shoe to the system itself, which allows us to monitor our jog. As all the trees of the city are fitted with RFID chips, we can monitor their state.

Bernard Benhamou appeals to our conscience. A RFID chip is virtually immortal; therefore, he demands a "right to silence of the chips" or, in other terms, the possession of deactivation systems that enable us to establish rights to silence. On our mobile phones, we connect certain sensors that provide information about what use we make of them.

The identity of men within these networks becomes, as a consequence, an important matter. We will use the distinction made by Marc Augé in relation to this topic.

9.5. *Marc Augé's four observations about identity*

Human identity – a subject we have already tackled – is of paramount importance. It is usually informed by our civil status, which is the kind of trace that stretches from an individual's birth right to his death. Marc Augé helps us grasp different kinds of identities by means of observations. This notion of civil status is separate from that of personal status, which represents data such as location or body temperature. Digital identity transcends civil status.

First observation: "Identity, whether individual or collective, is always in relation to another and interpersonal…identity is the product of an incessant negotiation" [AUG 08, p. 83]. Identity is fundamentally linked with relation. This is why, as we have already pointed out, the concept of identity is used in the domain of information and communication sciences.

Second observation: "The analysis of the logic and mechanisms of alienation is one thing, but the processes that they structure are a different matter. Living cultures are cultures in constant movement, which accept change and contact…living cultures are shifting groups subject to the tensions and forces of history" [AUG 08].

Third observation: "No culture is inherently egalitarian; each institutes its own hierarchies".

Fourth observation: "Multiculturalism, in order to rise above the contradiction between culture and universalism, should not define itself as the coexistence of cultural monads proclaimed equal in dignity, but as the ubiquitous possibility of the individual to cross into different cultural universes" [AUG 08, p. 84].

What we find interesting in Marc Augé's approach is how identity leads back to culture. As a consequence, our new individual with his prostheses and implants can be moved back to a cultural dimension i.e. the one of the new society in which disabilities no longer exist.

9.6. *Identity, individual and culture*

"The idea of individual remains subversive insofar as it means that the world is born and dies with me. All kinds of cultures built themselves up against this solipsism, which constitutes their strength, since Otherness is at the core of identity; individual identity can be defined, conceived, and lived only in relation to those of others. Conversely, this social sense vanishes if the individual dissolves in conformism, likeness, and orthodoxy. The individual finds his fulfillment in solidarity, but we also know that this gratification, in its highest forms (love, friendship), needs no institutional framework" [AUG 08, p. 141].

The debate sparked by Augé is twofold. The individual simultaneously defines himself through his difference from others, but also by his integration into the ensemble of the social system. Therefore, we can see the difficulties involved in building these famous social networks, where we have to hope both that the individual will reveal his membership to groups and that his mobile phone will help him make some connections, despite inevitably forcing him to assert his difference through the content he will create.

9.6.1. *The new santacteur model*

A new model that separates universality from individuality allows us to buy a life, citizenship, identity, nearly as if we were in a video game. In the previous sections, we mentioned two phenomena: the creation of communities and the significance of neutral places. These notions lead us to another important factor related to technologies. In video games or on mobile phones, we see the emergence of new possibilities: the purchase of life, citizenship or identity. The term "purchase" is perhaps a bit excessive, but it corresponds to one of the individual's existential needs.

9.6.2. *The purchase of life or perpetual rebirth*

Both Avital Ronell and Joël de Rosnay deal with the topic of perpetual rebirth. We are not talking about the purchase of an actual life, but rather the kind of purchase we make in video games when the villain has eaten our pac-man. Men are reborn in video games, not in real life. We are still following the logic of immortality that we pointed out when analyzing identity.

Joël de Rosnay starts with the idea of rebellion and the emergence of a "pronetariat" [ROS 06]. Industrial and economic models are changing. Our current industrial model is based on economies of scale: there is a concentration of means of production and mass distribution, supported by advertising and its passive consumers. The new emerging media present a decentralized production. Production is becoming widespread. This is why Joël de Rosnay uses the term "pronetariat", which is "bound to be born"; "proletarians" are associated with the capitalist system. Like capitalists, these new actors own their means of production. They use "empowerment tools". They have exchanged digital content, "user-generated content", music, texts and images. The advent of exchange globalization is becoming a reality. Joël de Rosnay is therefore in line with the logic of rebirth, in which different beings are going to be affected by co-education and the exchange of skills. This situation involves both the *santacteurs* that find out about their disease on user-built websites and those who take care to get tested regularly, particularly diabetics on insulin therapy.

Avital Ronell's perspective is more unique. She does not hesitate to talk about a mother call. "A certain Oedipedagogy is taking shape here – the restoration of contact is in the making, initiated by a mother whose navel, in Joycean terms, would emit signals" [RON 06, p. 31]. Ronell regards the navel as a third, if blind, eye. This leads us back to the history of Oedipus, Laius and Jocasta's son, upon whom many ancient tragedies like Sophocles' *Theban Plays* are based. Let us recall that by "Jocasta complex", we designate the mother's libidinous fixation on her son. Further on, the same author adds that Jocasta's call is "the way she secretly calls the shots and her responsibility", then that "once made, the call indicates the mother as *aufgehoben*[10], picked up, preserved, and canned" [RON 06, p. 34]. Avital Ronell describes this singular product in terms of the mode of communication that it creates. "Writing was in its origin the voice of an absent person; and the dwelling house was a substitute for the mother's womb, the first lodging, for which in all likelihood man still longs, and in which he was safe and felt at ease" [RON 06, p. 74]. This naturally leads us to the theme of rebirth: a call involves rebirth, at least emotional. Similarly, "the call of conscience possesses the character of a telephone call" [RON 06, p. 43].

Avital Ronell sees this perpetual rebirth as a cycle. "Returning in general, disappearance/reappearance – what was earlier cited as a relation to absence – transferring oneself into an object or placing a call to the non-there, and identifying with it: these motion to the events described by Freud in the now famous *fort/da* analysis" [RON 06, p. 71]. An individual appears in an application, and then he disappears. We unfriend people and they are therefore dead to us, but, later on, they will be born again, thanks to a re-integration into the unpredictable origins. Thus, we have to focus on the existence of "digital death" or on donating after death, when our organs are re-used for transplants.

9.7. The purchase of citizenship

At this stage, we have to focus on the notion of purchase of citizenship. Rather than belonging to a group, we abide by rules and norms often accepted as true but not written, which turn us into "citizens" in the sense of belonging to the Greek polis.

10 A German word that literally means "lifted up".

According to Albert Jacquard: "The city constitutes one of the stages of this progressive insertion into a set and of this interlocking of increasingly larger groups: associations, parties, nations. Being a citizen means being aware of this belonging". According to him, this form of citizenship is neither inconsistent nor abstract.

It "manifests itself, among others, in those lists that include my name, from alumni's yearbooks to lists of active members, from my birth certificate at the register office to my electoral registration. Each of these lists expresses a form of citizenship. If our name is in the telephone directory, we are already playing the game of citizenship" [JAC 05, p. 39]. For certain diseases, we buy our citizenship through registration, for example, to cancer registries or to the cancer files kept by modern healthcare systems.

This citizenship approach put forward by Jacquard has been compared to Ferraris' notion of registration, but it significantly transcends it.

Jacquard goes beyond it by introducing the concept of social link and the notion of autonomy. He mentions Paul Ricœur and the Enlightenment in order to define citizenship as the victory of rationality upon irrationality, which is revealed by its inhumanity. "Learning what citizenship means involves becoming aware of others' needs so that we can become ourselves" [JAC 05, p. 47]. Once again, the concept of link is of paramount importance.

10

Trust

In the traditional forms of industrialization, trust often involves branding and labeling. Labels allow us to justify a statistic quality and the abidance to a procedure or norm. Trust is also a tool of economic effectiveness. This is the reason why liberal economy replaces interpersonal trust with the trust in rules, which is an exchange that allows us, among other things, to trust strangers. This is also why the notion of a trusted third party has emerged. This kind of "industrial" trust is based on reducing the number of mistakes as much as possible and abiding by certain rules. In the health industry, the matter is evidently more complicated.

10.1. The source of trust

It is commonly accepted that trust is linked to a decrease in the number of faults and mistakes. We saw how we tried to reduce the number of undesirable medical events and how this constituted the source of health industrialization.

10.1.1. *Failures versus mistakes*

Failure occurs when one does not follow established and well-known rules, whether ethical or moral, i.e. the rules of the game or protocol rules, procedures, norms or regulations. Each successful industrialization involves the expression of these rules. In the medical field, however, what can we say about a doctor who is knowledgeable and skillful enough to be allowed to disregard the protocol – which would have definitely killed the patient – and whose behavior actually ended up saving him?

On the contrary, the notion of mistake involves getting something wrong and not knowing where to go. It is a specific kind of fault that we associate with an absence of purpose or vision. Mistakes occasion catastrophes, worst-case scenarios or, at best, a disruption.

Recent laws in the medical field have made a clear distinction between the two approaches. The Kouchner Law, back in 2002, introduced the concept of therapeutic hazard.

10.1.2. *Screening versus medical diagnosis*

Trust is often linked to results and to their nature. Therefore, the question of screening arises. It is possible to carry out a screening for cancer, which is then in a preclinical condition, whereas medical diagnosis certainly deals with clinical phases. No one can say whether cancer screening will become clinical or not. Doctors therefore adopt the precautionary principle, which makes recovery completely arbitrary and renders the marker represented by post-diagnosis survival time meaningless.

The concept of survival time after finding out about the disease raises at this stage the problem of defining a starting point. Does it consist of the appearance of the first cancer cell, in the screening performed geared to this or that method, or in the clinical diagnosis? This situation not only confuses patients, but is also puzzling from a public point of view in relation to the usefulness of this or that screening solution.

10.1.3. *Looking for effectiveness*

Standard practices lead to faults associated with the search for effectiveness. At this stage, we have to mention the question raised by Perriault: "Is the notion of logic of usage integral part of the concept of individual as a whole or can the two be separated?". This is also what Jauréguiberry and Proulx propose:

> "We prioritize motivations in terms of effectiveness, profit and viability, leaving out factors of emotional or relational

> affection, belonging or identity. The choice made in terms of communication technologies is mainly presented as the result of an economic way of reasoning. It is a matter of managing emergencies, making down-time profitable, optimizing tasks in real time or, in short, being efficient" [JAU 11].

Therefore, a company that reaches its goals is effective. It pivots around the mediation of rationality, which is made of relations and individuals. From a functionalist point of view, society is a sum of elements, namely the individuals, relations and institutions that make it up. It comes before these elements, functions in a specific way and makes use of tools. By means of tools, society has at its disposal some specific dynamics that enable it to solve situations of instability. It has a natural tendency to resist change. All of this results in a form of rationality that is often described only afterwards. This notion of effectiveness or accomplishment, which could present itself in the medical field as the treatment of every patient, should not be confused with that of efficiency, which consists of treating while spending as little as possible.

10.1.4. *Errors in judgment*

Several errors in judgment or behavior are used, often too simplistically, as answers. The first consists of the principle of scapegoating, which has been practiced since ancient times and involves preventing the person who has made a mistake from being harmful. Let us point out that in countries like France, it is one of the significant activities in which professional boards engage.

Another wrong approach consists of attempting to rectify the mistake. The principle applied in this case is the so-called "plaster on a wooden leg". The rectification performed by means of another system necessarily creates new types of errors. The robotization of hospital pharmacies undoubtedly leads to the creation of safer medication systems, but it also generates, in its wake, other malfunctions such as medication arriving too late or the impossibility of using similar medications as a replacement when there is a shortage of certain substances. These malfunctions cause people to die.

Rating, or benchmarking, errors is the most blatant of mistakes. We should rank hospitals so that we could find out which are most liable to make mistakes. This idea would be completely pointless: at best, ranking would enable us to find out levels of mismanaged competences and procedures resulting from an act of incompetence. The element at risk is by definition unpredictable and statistically affects every establishment in the same way. Otherwise, it would not be a risk element! In one of the previous chapters, we mentioned some raw data. Let us now put it in perspective! In France, an accident takes place six times per 1,000 days of hospitalization, or weekly for a 20-bed hospital department.

Another source of error consists of the idea of registers. Registers provide information on deaths and accidents, but give us no clue on how to prevent them. Moreover, they lead to excellent correlations which are too quickly turned into causalities that will turn out to be completely wrong! Of course, deplorable working conditions are not an element of longevity and do constitute aggravating factors, but they are not the cause of certain diseases like some types of cancer.

The last kind of mistake consists of believing that by abiding by the established procedures, we will manage to solve these problems. On the contrary, we are locking ourselves into the error mentality by thinking that we are doing the right thing without managing to adjust.

10.1.5. *The stages of competence*

The degree of autonomy or competence that we attribute to the subject as he is described in his activity of consumption is constantly being questioned. In a rather old work on this topic, Don Slater observes that academic descriptions of the consumers' life swing between two extreme perspectives:

– a cultural dupe or dope;

– a hero of modernity [SLA 97, p. 33].

This approach can be perfectly applied to the health industry, where men are either dupes before an omnipotent doctor or individuals wishing to be treated with the latest technologies. However, the matter is not that simple and binary. This notion also plays a significant role in the relation between doctor and patient. The doctor's authority was based on these competences and it was the element that allowed him to dominate the relation with his patient.

According to Zygmunt Bauman, this approach also divides consumers into two groups. We could easily replace the term "consumers" with the words "patients" or "sick people".

"At the first extreme, consumers are presented as anything but autonomous individuals: they are shown instead to be hoodwinked by fraudulent promises, enticed, seduced, pushed and otherwise maneuvered by blatant or surreptitious, but invariably extraneous, pressures". These are represented by the appeal of medical advertisements, who affect patients as much as doctors.

At the other extreme, "the consumer encapsulates all the virtues for which modernity wishes to be praised – like rationality, robust autonomy, capacity for self-definition and rugged self-assertion". According to Bauman, this does not result in global sharing carried out between "things *to be chosen* and their *choosers*", but on the contrary in the "*blurring*, and ultimately the *effacing* of the divisions listed above". The most striking feature of the consumer society is "the *transformation of consumers into commodities*; or rather their dissolution into the sea of commodities" [BAU 08, p. 22].

Author	Theory in favor	Theory against
Don Slater	Hero of modernity	Cultural dupes
Bauman	Encapsulating the virtues of modernity	Anything but autonomous individuals

Table 10.1. *Perspectives on the consumer society*

The differences pointed out by these two authors are relevant to the health field.

This notion of modernity was the constant preoccupation of the Frankfurt School. This school of thought was pessimistic about the fate of modernity essentially because of a global phenomenon that characterizes it, namely authoritarianism, which was then starting to take different shapes. These involve, in the medical field, the directions that we need to follow. We will rarely employ this notion, since we prefer instead the term "contemporaneity", which we will use later on.

10.1.6. *The significance of relations and links*

This interconnection is, as we pointed out, very important.

The process of establishing connections is part of the dynamics of technological progress. Data processing must be "located in history", i.e. in the successful 20th Century, which lays the foundation for relational data based on a two-dimensional system. This success involved the possibility of connecting dispersed pieces of information of the same kind. This model presupposed a certain degree of stability.

The advent of technology changes therefore the link between patient and doctor. As a result, we shift from an authoritarian relation, which gives rise to trust, to a reputation-centric system. In other words, the pair authority-trust is replaced by the pair reputation-trust. If this event takes place, it contradicts the philosophy proposed by the Frankfurt School.

10.2. A resilient approach

DEFINITION.– *Resilience is the ability to adapt in the face of variability by constantly predicting mistakes.*

10.2.1. *Why resilience?*

Resilience, as a notion, tries to solve the current situation where laws, rules and procedures have not been able to predict a particular or unexpected event. These situations are all the more frequent in the health industry since we abandoned the concept of competence as source of authority and trust in favor of reputation, which can be "created" in other ways.

Several studies, especially those conducted by the American Institute of Medicine, have shown that there was no connection between accidents and the implementation of procedures. Certain case studies show that the failure to comply with protocols has saved some patients' lives, in particular when it comes to vaccinal approaches on patients affected by cancers.

Resilience has to be linked to competence, skills and talent in order to manage the unexpected. The other element that can facilitate this task is the patient himself, who can turn into a *santacteur*.

Resilience is hard to assess. Some estimates about health-related Internet searches in the United States can be interesting and help us to understand it.

Some estimates about the American population
Out of a total population of 310 million people, around 100 million Americans use the Internet to look for information, essentially in terms of prescriptions and medications. Among these 100 million e-patients, 85.6 million claim to have doubts about healthcare professionals' opinions, diagnoses and the role of medications, which is what justifies their actions.
61% of adults look online for information on their health. This estimate goes up to 88% for users of broadband networks. In total, 57% of them are men and 64% are women.
51% of patients affected by chronic diseases have looked for information on the Internet.
42% of people claim to help friends of family members look for information online.
66% of patients who use the Internet state that the information obtained has modified their behavior in relation to their treatment.
The estimate for people looking for information online increase with education and income levels, while dropping with age.
66% of them use search engines, an estimate that goes up to 77% for people who have been recently diagnosed. However, only 17% use their mobile phone. 35% post information on social media, 60% wish to receive information this way, 38% are willing to discuss health on social media, 51% of them claim to be in favor of the social media of charitable organizations and 37% state to support the media of pharmaceutical companies. However, only 12% wish to see advertisements, including those for prevention.
Source: National Institute of Health, USA, 2013, most of the surveys date back to 2010.

Box 10.1. *Internet searches for health matters*

10.2.2. *The HACCP approach as a solution*

The concept of resilience allows us to make use of the industrial approach called TQM (Total Quality Management) in the health industry. Engineers have employed an approach called CCP (Critical Control Points). It is a method devised by NASA in the 1960s, which has evolved into the HACCP approach (Hazard Analysis and Critical Control Points). It consists of seven principles.

Principle 1: Conduct a hazard analysis
Principle 2: Identify the critical control points
Principle 3: Establish critical limits for each critical control point
Principle 4: Establish CCP monitoring requirements
Principle 5: Establish corrective actions
Principle 6: Establish procedures for verifying that the HACCP system is working properly
Principle 7: Establish recordkeeping procedures

Box 10.2. *Basic principles of the HACCP method*

Hazard analyses are systems that determine safety risks and identify the preventive measures that can be implemented to reduce them. A risk is defined in terms of biological, chemical and physical properties or as a procedure that may render an action unsafe. Hazard analysis has to go further than mere clinical analyses and the verification of medical devices.

A critical control point is a step, an action, a stage or a procedure within a global care process on which we can exert control. As a consequence, risks can be predicted, eliminated or reduced to a rate acceptable for that level. However, we should still be able to identify these control points, which only experts are qualified to do.

The third principle is certainly the trickiest to implement in the health industry. It is a matter of establishing critical limits, namely the maximum or minimum values for which a physical, biological or chemical risk has to be managed to a critical control point. The risk has to be prevented, eliminated or reduced to an acceptable rate. Sometimes we are aware of it but, in certain cases, it remains very difficult to take the necessary steps since we need infrastructures. Strokes (cardiovascular accidents) perfectly illustrate this point. Doctors have known for a long time that the first half hour is critical, but they still need a neurologist and the means necessary to carry out the thrombolysis within this time frame. The Alzheimer Plan in France has allowed some solutions to be implemented in certain regions. Finally, monitoring activities are required to make sure that the process is under control at every critical control point. Each monitoring procedure and its frequency could be listed as part of a plan. This is the application of the fourth principle.

When the monitoring process indicates a deviation from one of the established critical points, or whenever a major incident happens (such as a power outage), the fifth principle points out the actions that need to be taken. It is always a complex task to identify the corrective actions that have to be performed if a critical limit is not respected. Currently, as for medical robotics and certain medical devices, this action is established only through clinical trials, which in this case may not be enough. Corrective actions have to be predicted so that we can make sure that nothing harmful happens.

Verification ensures that industrialization does what it was devised to do. In other words, industrialization is successful if it ensures the performance of a safe action. This constitutes the sixth principle. Both healthcare establishments and independent professionals will be required to have their own plans validated. Control bodies will have to approve these plans beforehand. They will go through them and verify their compliance with healthcare protocols, administrative and legal rules, etc. Verification also ensures that the plan is suitable and therefore guarantees that its effects will correspond to the established ones. We can include in these verification procedures an analysis of critical values and of sampling and analytic methods. This process can lead us to find scientific "evidence" that can verify the critical points defined at the beginning.

As for the seventh principle, it consists of having and keeping certain documents at our disposal, as we do for risky industrial environments (Seveso sites) and air transport. The monitoring of critical control points, verification activities and "deviation" management is documented.

Pharmaceutical industries, which usually manage Seveso sites, are certainly in the best position in this domain. On the other hand, we should point out that the implementation of such methods in the health industry meets with opposition, whereas the American Academy of Sciences started promoting this type of approach in terms of food safety back in 1985 and it has been used in civil aviation since the 1960s.

10.3. The creative man

One of the new elements that characterize the modern health industry is how men take control of themselves. The shift to a type of medicine based on reputation is certainly the source of the decreased number of people blindly trusting healthcare professionals, which translates into the desire to take control of ourselves.

10.3.1. *Man, between user and producer?*

At this stage, we have to examine two points that were recently integrated into this debate: on one hand the body as digital and technological presence and, on the other, human needs as reclaimed by the technological healthcare system.

The super-system man/health is characterized by a predictable result: health. This implies fewer deaths, fewer handicaps and a more pleasant life. Consequently, our only solution – as we have already remarked in another work on e-health and telemedicine – consists of focusing on man. By "man" we mean here an individual with no distinctions in terms of sex, but we distinguish this concept from the notion of patient.

One of the principal ideas consists of having the *santacteur* gain competences, helping him produce data that will make future diagnoses easier, and above all giving him a better life straight away. Nowadays, discourses on users' data production, which translates into knowledge production, go well beyond websites like Wikipedia or Facebook. The individual may play a part, but technologies can also help him contribute.

10.3.2. *The individual as an actor*

A communitarian vision of the fabrication of the tools employed to capture health-related data – be it a sugar-testing pen device, tablets or mobile phones – involves some thinking. Before technologies become invasive, the user is required to generate content. This goes beyond the mere fact of revealing our health state on Facebook and transcends even more the debate on transparency. Users are now creating data and not a simple vocal exchange, as is the case for the doctor-patient meeting during the appointment.

10.3.3. *A sense of body*

A new approach, advanced by Richard Harper[1], focuses on the sense of body in terms of digital presence: we have to show people that we own one. According to him, several changes have taken place, since men are simultaneously becoming recipients and producers of information. In the world of communication, we should not forget that we have shifted from physical constraints to digital freedom, from geographic specificity to digital omniscience and from editorial content to User Generated Content (UGC).

According to Richard Harper, this development is key to understanding social networks, which requires an adjustment of communication systems to bodies. He regards these processes as "relations for adaptation". In this information-centric world the individual, who was formerly a programmer, has now become a player. If beforehand he was relegated to the background, he is now becoming fictional. The new individual must share and use a piece of information as comfortably as possible. How hypertext has modified the notion of "telling" is just one example.

The theorists of social networks are well aware of that. It is not exactly what they imagined, since transaction costs dominate every aspect. Current social networks are above all tools for "staring" at someone. In these networks, we claim for ourselves a certain identity which defines our interactions. Harper points out the paradox between the notion of fixed and timeless identity and the individual, who is actually a set of self-regulated systems – one of which is health – within the time that represents them.

This distinction is quite old but still pertinent. In our case, it is expressed in the production of content, which is becoming utilitarian in relation to the life of the individual. This leads us to the definition of those approaches related to mobile phone use, like the GEMS one which we will shortly describe. These approaches are more modern than the one put forward by

1 This section is drawn from Richard Harper's conference (Professor of Socio-Digital Systems-Researcher, Socio-Digital Systems, Microsoft Research) called *Les entretiens du nouveau monde industriel*, held on 3 and 4 October 2008 at the Centre Pompidou in Paris. It was organized jointly by the Institut de Recherche et d'Innovation (IRI)/Centre Pompidou, the competitive cluster Cap Digital and the ENSCI-Les Ateliers (Ecole Nationale Supérieure de Création Industrielle).

Philippe Mallein and Yves Toussaint [MAL 94], which is based on the articulation of the analytic categories of usage, representation practices and context.

10.3.4. *The theory of affordance*

An affordance theory was advanced by Thierry Bardini. According to him, it is a matter of grasping the relationship between object and user and measuring what the user gives to his object, which is a combination of actions, contents and events. Thierry Bardini transcends therefore Ferraris and Akrich's concept of registration by introducing a relation between the object and its user. Thus, the user associates physically and concretely with his technical tool, going beyond a mere relation of textual or vocal exchange. According to him, this affordance also introduces the user's perception. Therefore, the final result derives as much from the way the object works as from the user's expectations and their mutual "perspective" [BAR 96].

10.4. The dynamic approach (GEMS)

The GEMS approach was proposed by Lehikoinen, Aaltonen, Husskonen and Salminen in 2007 [LEH 07]. It is based on the dynamics of community excitability. GEMS stands for Get, Enjoy, Maintain and Share.

10.4.1. *A description of the GEMS approach*

This description can be found in Figure 10.1.

This model shares some aspects with the "analysis of the lifecycle", especially the one of digital information proposed by Gilliland-Swetland [GIL 06].

In the GET phase, the user has to take action. A picture must be taken. A document must be written. A purchase must be made. During this phase, technological metadata can easily settle.

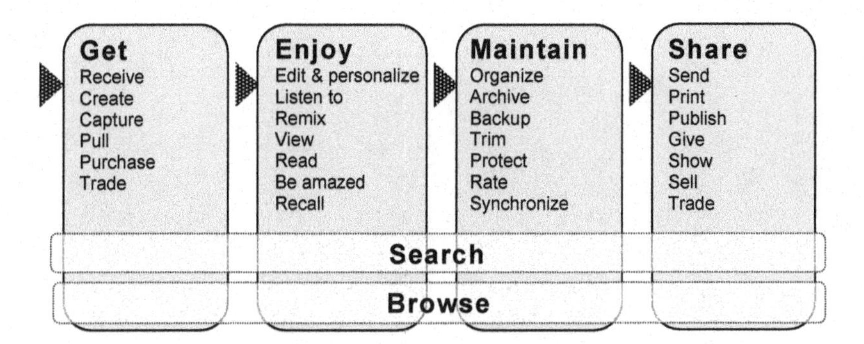

Figure 10.1. *GEMS model*

The ENJOY phase consists of the development of these acts and represents a key point in the whole system. Anything necessary for the internet user to optimize this phase is included.

The MAINTAIN phase is constituted by the global management of the system. Technological rules play a significant part in this field. Currently, the majority of solutions are closed-source software, with the exception of Google, which has recently come up with a free one.

The SHARE phase enables us to access the result of these actions. The main difficulty at the moment has to do with browsers and terminal interfaces

11

The Social Reality

11.1. Singularist and transhumanist theories

Machines are intruding upon our daily lives. People are not actually aware of how much they are coming to rely on them. They also shape our behaviors and some people's way of thinking. These tools are thought to make us more efficient or, in any case, this is how they are sold to us. According to some, domestic technology should have made us save time. For others, the aim is to set us free by helping communication in a society where moving is becoming the rule.

Prostheses, orthotics, bionic men, AIs or robots are becoming more and more autonomous. Research, the aims of which may be more or less known, leads us to a new era sometimes called "transhumanism" or nicknamed "theory of technological singularity". It is not only a school of thought, but also a matter of new technologies linked to sciences and new subjects with the potential to modify crucially the physical and intellectual abilities of the human species. According to some, this could give rise to new types of life termed "posthuman". This chapter deals with these theories, compares them with our reality and attempts to locate them inside and within the process of industrialization. With respect to medical progress, skeptics are keen on extrapolating, whereas optimists are looking for solutions that, like switches, can bring about change. This is exactly what characterizes transhumanism or singularism.

11.2. DNA microarrays and their use

The presages of health can be seen in service companies operating in the domain of life sciences. This sector is still atomized.

Microarrays are the fundamental technology of the genome age. These slightly unique chips are sold by several vendors and have a wide range of functions. This technology is approved for taking types of RNA as well as other biological measurements. Despite everything, microarrays remain research tools. The main suppliers are: Agilent, Affymetric, Illumina and Nimblegen.

These RNA microarrays allow us to employ a method called RT-PCR or reverse transcription polymerase chain reaction. In very simplified terms, the method consists of building several copies of the part of RNA that we are searching for, amplifying them and then observing them using a fluorescent dye called ethidium bromide, phosphorescence or scintillography techniques.

IT companies employ these chips. Genedata provides IT analytic tools under the name of Expressionist. This kind of software works as a sieve. The company called Genedata provides a flexible, visual and intuitive platform that easily processes results for millions of measurements in a few minutes, regardless of the complexity of the experimental system (high throughput screening or HTS).

The best known applications are the search for "defective genes" and oncology. Companies such as Oncodesign have developed solutions in the fight against cancer on the basis of principles of transrational medicine. The technological platform of Chi-Mice© is regarded as an *in vivo* modelization. This solution deals with the development of models suitable for the selection of effective therapies against cancer. From the use of patient-produced xenografts to rodent models rebuilt with a human immune system, we can assess the medication taken by patients in clinical-appropriate conditions. It proposes the Chi-Mice platform. For that reason, it allows us to better take into account the pharmacological studies that employ human cancer models in order to optimize defense mechanisms against it and therefore devise better chemotherapy strategies.

11.3. The economic reality

To discuss about how much social security can pay is appropriate when talking about the industrialization of medicine, but the economic debate is actually more complex. Currently, care management is based on a cure model. Therefore, a prosthesis like glasses will be reimbursed only moderately, if at all. On the other hand, certain incurable chronic conditions have been reimbursed under the pressure of economic actors. This model will not be able to last and industrialization will be the solution.

In the short term, the actual question revolves around regulations, their coherence and their stability, since development times in the health industry will most likely remain long-term (about 10 years).

11.3.1. *Enquiring patients*

This model will have to involve patients in a more significant way. It will actually be necessary to turn the patient into the long-awaited *santacteur*. This *santacteur* will look for information, try to predict as far as possible, and involve several actors, some of whom do not currently exist in the healthcare chain.

11.3.2. *The economic logic*

The economic logic consists of the concept of service rendered, which is driving little by little the actors involved to incorporate medicine into a classic logic of service. Therefore, if the medical service rendered corresponds to actual progress for the patient, the price will naturally be higher. This entails that the proposed prices are becoming step by step economic aberrations or incentives that can turn out to be either effective or, on the contrary, inhibitors of progress. Prices regarded as an effective compass require us to have a global perspective on the cost entailed by taking charge of a patient, who may be affected by one or more diseases, present one or more symptoms and necessitate one or more treatments. This is certainly not the case in countries like France.

If a country does not accept to play this economic game and to keep abreast of the times, the consequence will be incontrovertible. Innovative companies will move elsewhere and set up their equipment abroad and a form of medical tourism will develop.

11.3.3. *The role of telemedicine*

Telemedicine is one of the key factors of this economic progress. As Pierre Simon, former relator on this subject and president of ANTEL (French National Association of Telemedicine), loves recalling: “15% to 20% of hospitalizations could be avoided thanks to telemedicine!”.

11.4. The new industrialized health

The new industrialized kind of health is characterized by some conceptual developments highlighted to avoid the shock evoked by the term “industrialization”.

11.4.1. *The four Ps approach*

Researchers at the National Health Institute in the United States have tried to describe as well as they could this approach based on the individual who is not necessarily sick or a patient, which is a revolution in itself.

The first action consists of tracing back the signs of the diseases to their earliest stage before the appearance of the first symptoms.

The second one consists of developing medical practice through the shift from a care process to a system of good-health management. Hence the concept of the four Ps, most fervently supported by Elias Zherhouni[1]. Medical progress will have to evolve around four key principles: predicting, personalizing, preventing or pre-empting and participating.

1 Elias Zerhouni’s conference held at the Collège de France as Bettencourt professor on 24 January 2011, on the theme of “main tendencies of biomedical research”.

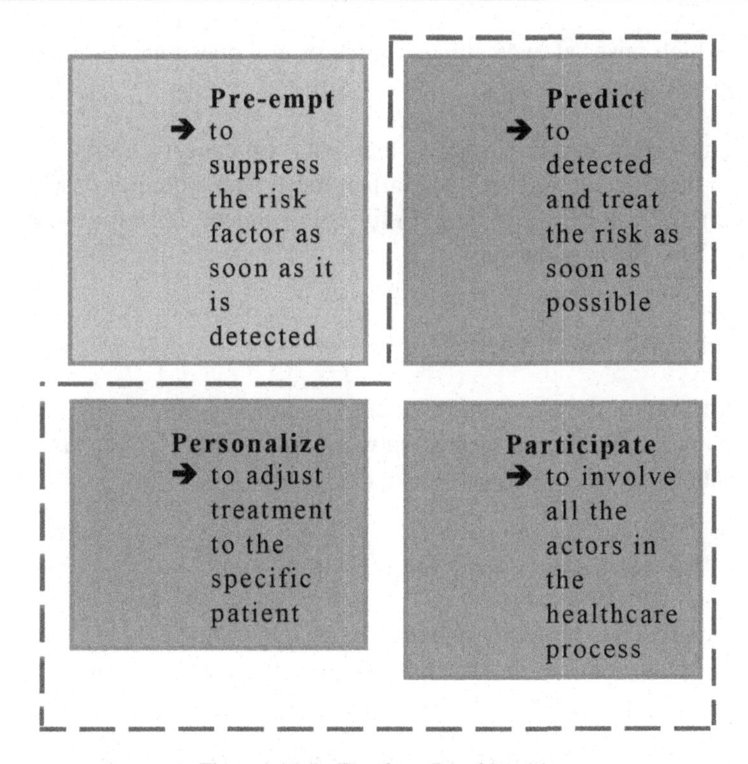

Figure 11.1. *The four Ps of health*

11.4.2. *Prevention or pre-emption*

The shift from prevention to pre-emption is often described as wishful thinking. How can people be prevented from smoking and drinking? Every kind of measure that resembles prohibition is unsuccessful.

11.4.3. *Personalized medicine*

Personalized medicine determines treatment on the basis of tests carried out on the individual. This is why we talk about "companion diagnostics". This allows us to target our treatments. In 2012, only 1% of medications sold benefited from companion diagnostics, whereas experts estimate that 10% of them take advantage of them. Cancer, asthma, schizophrenia and Alzheimer's disease are all privileged fields. Personalized medicine consists of saving treatment only for those patients who are liable to produce positive effects, which raises at once ethical problems and questions in terms of equal access to healthcare.

Experts estimate, for instance, that 30% of cancers could be treated, thanks to these solutions that never affect more than a quarter of all patients. The estimated values would be 25% for colon cancer, 20% for breast cancer and only 12% for lung cancer.

11.5. Debating the model

The debate on the economic model of health is certainly the core of this reflection and nowadays it swings between two extremes. We can analyze it as a debate between an engineer and a doctor.

11.5.1. *The engineer's approach*

This approach is perfectly represented by the white paper proposed by the Syntec Numérique (French trade union part of the digital industry).

11.5.2. *The range of possibilities of the SNITEM model*

Based on the analysis of international examples, the SNITEM model proposes different ranges of possibilities, described in the diagram below.

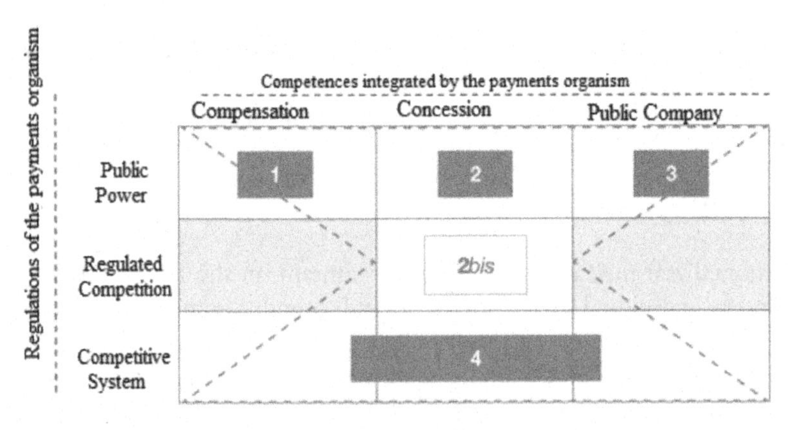

Figure 11.2. *Different possible approaches of the SNITEM model*

The four possible approaches are:

1) refund of telemedicine;

2) model of integrated care;

3) dedicated public healthcare network;

2a delegation of public services;

4) privatization of healthcare.

As the report reminds us, the main payments organism of the French healthcare system is determined by non-competitive public regulations. As a consequence, the Syntec is limited to the first three scenarios (1, 2 and 3). The other two (2a and 4) involve a change in the regulation system of the payments organism. This development does not seem likely to become a reality; however, it could be the result of major financial constraints in the future. The question will then revolve around establishing whether this privatization is competitive or organized.

11.5.3. *The doctor's approach: ANTEL replies*

The French National Association of Telemedicine, which groups together mainly doctors, has reacted to this debate.

11.5.4. *The market approach*

The problem raised by personalized medicine is twofold. At first, policy makers focused on reducing the size of the market created by personalized medicine. Afterwards, the tricky aspect consisted of the collaboration between two types of companies, namely those who produce the solution and those who perform the diagnosis. Thus, two kinds of establishments were set up: either research centers, such as the one in Pernzberg (Germany), or start-ups that have found this niche with the co-operation of other actors.

Disease	Medication	Companion diagnostic
Lung cancer	Xalcori (Pfizer)	Roche Diagnostic
Colon cancer	Erbitux (Merck Serono)	Qiagen

Table 11.1. *Examples of medications and companion diagnostics*

Even if the medical diagnosis market remains weak (it allegedly reached 500 million dollars in 2012), it remains indispensable for gaining access to the target molecules market, which seems incredibly large: 6 billion dollars for Roche's Herceptin, used for breast cancers, 1 billion for Novartis' Tasigna, used for leukemia, and 360 million for Amgen's Vertibix, used for colon cancers.

11.5.5. *Medical tourism*

India is the global leader of medical tourism, drawing in patients from abroad who are looking for medical treatments such as the insemination of frozen embryos on surrogate mothers. There is an society between countries involved in medical tourism, and annual conferences on this topic are held in the countries that are in favor.

12

The Technologies that Could Change Everything

The technologies with the potential to "change everything" have to do with substitution or switching approaches.

12.1. Biotechnology

12.1.1. *Transplants, cells and cell therapy*

Transplants and stem cells are emergent technologies that represent types of industrialization. Transplants can also be artificial, as is the case for hearts and pancreases.

An American researcher called George Dvorsky published an interesting article. By transplanting human glial cells into rodents' brains, neurologists could "suddenly" enhance some of their cognitive abilities. These enhancements involved improvements in terms of memory, learning and adaptability. The cells extracted from other rodents had no effect. These solutions have been dismissed for a long time as mere support structures for feeding and managing neurons. It was recently proved that these cells can have significant effects on information processing and may play a part in the development of a disease. These results come as good news for those who are in favor of extending our life span in terms of active living and good health. These products are essential to the increase in or return of types of human intelligence. This gives us hope in the fight against degenerative diseases, in particular Alzheimer's and Parkinson's disease.

A company called Cellectics offers cell therapies, especially for cases of diabetes and leukemia. It is a technological platform centered on T cells which allows us to treat type 1 diabetes. Its aim is to engineer pancreatic cells that can be used on an industrial level. Let us recall that pancreatic cells produce insulin. Clinical trials should start around 2015.

Another French firm, Gensight, focuses on gene therapies for eye diseases, especially Leber's neuropathy and retinitis pigmentosa. The former disease is mitochondrial – and consequently genetic – passed on exclusively by females and by now diagnosable at birth, which makes it a classic disease.

12.1.2. *The applications of biometry*

The use of biometric data is based on the hypothesis that it enables us to identify a person from the biological characteristics of a part of his or her body. Fingerprints were the first kind of biometric data used by the police. Then research focused on the iris, the vein system, hand outlines and facial features. Finally, DNA genetics emerged as a foolproof weapon.

Biometry used against crime to prevent frauds and identity theft is now a reality. Some researchers have unearthed the humoral theory and hope to apply biometry to medicine. The head set-ups used to monitor the brain analyze variations in the venous system.

The use of biometric data raises several problems. The first has to do with identification, which is now made possible. Every French passport issued after 2009 has been biometric. In October 2011, a decision of the Council of State required the suppression of 40 million fingerprints, so that only two fingerprints out of the eight initially taken were kept. On the other hand, the Indian Aadhaar system, as part of a census conducted on the whole population of 1.3 billion people, managed to provide everyone with a biometric ID number. Moreover, the ways used for carrying out recording and experimenting seem to be important. Another problem is more ethical: biometry can be used not only to control access to buildings or facilities, but also to make online purchases and trade safer. The boundary can soon be crossed if we prevent diabetics from walking in a restaurant or buying products containing sugar online. It is a matter of misuses. Another issue has to do with reliability and what it entails. What is the risk of falsification?

How can we ensure the protection of biometric data? The last question raises the problem of databases and inscription.

12.2. Energy and material technologies

12.2.1. *Energy*

The rapid drop in the prices of renewable energy and in the combined cost necessary to store it constitutes another "switching" approach. Even if solar energy cannot compete with the price of oil, natural gas and electricity yet, its technologies are starting to be used to supply artificial elements incorporated into the human body.

12.3. Materials

Among the new materials that will revolutionize the health industry, we should mention zeolites, graphene and synthetic biomaterials.

12.3.1. *Artificial zeolites*

Zeolites, literally "boiling stones", are natural minerals with the potential to prevent gases from escaping: matter becomes a trap for matter.

New materials, called MIL (Materials of Institut Lavoisier), have certain properties that can be innovatively applied to the health industry. They are named after the laboratory where they were discovered. These materials are solids called "porous".

They have three unique features:

– a 3D structure is identifiable and known. In order to represent them on an atomic level, we have to imagine a block of cheese with several huge holes recurring regularly and steadily;

– this frame is linked to a reactive internal surface;

– internal cavities can lock up molecules. As nature abhors a vacuum, these cavities are certainly there to be filled. All the research consists of finding a way to empty them out. Therefore, it is possible to trap and store gases like hydrogen, oxygen and carbon dioxide.

These new materials offer a wide range of applications. They can be used as molecular sieves, thus separating the molecules present in blood, or help us to create new tests. They can also trap gases and could therefore be used to create artificial lungs. They also act as catalysts and ion exchangers, which researchers conceive as livers, intestines or kidneys. One of these materials is called MIL-101. It has an interesting property: a liter of it can store up to 400 liters of carbon dioxide at 25°C without increasing in volume, since the gas is compressed into the pores of the material. Other compounds can swell up to 300% their original size, before returning to their initial state. Iron-based MILs turn out to be particularly promising nanocarriers that could transport medications into our organism up to the target organs or cells, and thus constitute weapons against cancer cells.

Computer simulation enables us to understand the performances of the materials. One of the areas of research consists of imagining and conceiving these 3D structures. We then move on to the next step, which leads to the synthesis process, before finding out the possible applications.

12.3.2. *Graphene*

Graphene is characterized as a molecular structure similar to a layer of carbon, which is associated with graphite in a special form comparable to a heap of organic sludge. The list of current applications, from electronics to biochemistry, is huge. A recent piece of news in particular gives us reason to hope: the use of graphene for the creation of ultra-thin membranes.

Engineers at Lockheed Martin have just announced the development of a recently developed filter used to desalinate seawater. It could reduce the energy costs involved in desalination by 99%. Other filters with biological uses are being developed, which gives us reason to hope that we will be able to replace membranes in the human body. The solution typically involves a membrane sheet (TFC) made of a thin polyamide layer stacked on top of a porous polysulfone one. The thickness of these membranes requires high pressure in order for water to be pushed through. The "Perforene" filter engineered by Lockheed Martin is therefore built from graphene layers. As for its application in the health industry, we will have to solve other problems, such as tearing and biocompatibility.

12.3.3. *Synthetic biology*

Synthetic biology has become the art of fabricating beings. The creation of new forms of life is about to become a reality. Synthetic biology is an actual kind of biological engineering that will eventually enable us to create living organisms. At the moment, the tools used for computer-based assisted biological reproduction allow us to devise biological functions that cannot be found in nature. These are new materials that will be integrated into more complex products.

This developing technology lures us with several promises: stretching the limits of our knowledge on the origin of life and the transition between being and inorganic, but also developing numerous industrial applications like biofuels, medications, biomaterials, etc.

If the purpose declared is always the same, two main approaches seem to have the potential to make us reach our goal:

– a strategy based on building basic BioBricks designed to be assembled;

– engineering simplified synthetic genomes.

Different perspectives clash in relation to these two approaches. The first one tries to build actual "living machines" which will not be patentable because of their organic nature. The development model will therefore consist of IT open source, with free access to knowledge. The second one looks for the patentability of the synthetic compounds obtained, with an openly commercial purpose.

This artificialization of beings raises several problems in terms of ethics, besides this debate on patentability. The core of the debate is safety in relation to situations of voluntary dissemination or accidental rejection. Some propose strict regulations and control, even interdiction, as long as we cannot predict the performances of these synthetic organisms at large.

12.3.4. *A new creative dimension: 3D printing*

Being able to make objects quickly and meeting the demand has become a reality thanks to 3D printing. These technologies nowadays allow us to fabricate an object modeled in 3D. Therefore, it is possible to create frameworks on which human flesh will be able to grow, and to give an anatomical shape to a heart or liver, etc. Other applications, such as the creation of synthetic bones for the replacement of a natural one – whether worn out or broken – are also possible. 3D data can be obtained:

– by scanning an existing limb;

– by making use of medical imaging and separation-reconstruction software;

– by using computer-aided technologies (CAx).

Additive methods work by adding matter. Thus, the object is built layer by layer. It can also combine several materials, which is sometimes very useful in the medical field. Subtractive methods consist of removing layers, which is what classic machine tools do (plastic, wax, metal, etc.). These solutions have an economic advantage in terms of customized, and therefore small-scale, production. They also allow us to devise objects that would be very tricky to fabricate with classic equipment. The possibility of engineering an artificial bone that can later be reintroduced in the human body is worth a mention.

12.4. The "nano" world

The "nanometer" dimension covers different settings, all characterized by potential hybridization, since electronics work on a molecular scale.

12.4.1. *Nanosystems*

Different kinds of nanosystems are currently being developed.

It has been a few years since auricular implants allowed deaf people to perceive sounds. Some companies, like NXM in France, produce these devices and surgeons are already implanting them. Micro- or nanochips implanted and connected to the optic nerve or to the appropriate areas of the brain have already been explored as solutions for those who are visually impaired.

12.4.2. *Nanoproducts*

A product called NBTXR3, engineered by a company called Nanobiotix, can be considered as another example. Hafnium oxide nanoparticles allow us to strongly increase the power of radiotherapy by focusing it only on the cells affected. This company, set up in France in 2003, started its clinical trials in 2012 in conjunction with Pharma Engine, another company based in Taiwan.

12.5. Information technology

Information technology plays a role of paramount importance in the industrialization of health.

12.5.1. *Modeling life*

We will only provide the reader with a handful of examples.

Gap junctions are connections between astrocytes (a kind of glial cell) that also include astrocyte-neuron links. Gap junctions in neurons become standard synaptic connections. They create a sort of intercellular short circuit and enable the creation of high-speed networks for the propagation of signals in some areas of the brain, of the eyes, of the heart and of other parts of the body. Some gap junctions are sometimes called "electrical synapses". Currently, research is being carried out to engineer these junctions, either by computerizing them or by reconstructing equivalent biological devices.

12.5.2. *New digital tools*

New digital tools (mobiles and tablets together with their software) are used by children and even preschoolers. These new electronic formats help learn and think. Research shows that the "digital natives" of the new generation are less agile and spend less time multitasking. This "touchscreen" generation shows that the situation is not that simple. Electronic formats give new competences but, in this field, we do not know what is actually working and what is "lobotomized".

New mobile devices already come with chips that track users and monitor their movements. The latest ones can measure pressure, temperature and the level of humidity. Future devices will be fitted with software that follows our eye movements, which has already been used on screens for a few years. This paves the way for new applications and interfaces that require no keyboard, mouse or screen. Some will regard this aspect as an extra feature. However, eye movement monitoring can also present a dark more psychological aspect. Whatever happens, someone will be able to realize what we are looking at and therefore deduce behaviors and interests.

The new technology called "memristor" engineered by HP can create flash memory with the same speed of RAM memory and very high areal density. A cm^2 would be able to store 100 GB, whereas a cm^3 could store a petabyte (1,000 terabytes).

12.5.3. *Computing power*

Some of the new technologies will fill with enthusiasm people who are interested in the idea that sci-fi notions, such as approaches that increase our own intelligence and treat mental disorders, might become a reality. On the other hand, they will appall other people, such as those who hope they will soon be able to upload their minds to immortal robots.

Ray Kurzweil and other members of transhumanism are talking about "connectomes", which consist of simulating the number and locations of synapses that "light up" or "sparkle", thanks to ion transport between axons and dendrites, the 100 billion neurons and the trillion synapses of the human brain. This "lightning" can currently be seen thanks to the tools used for medical imaging such as MRI. Therefore, it is possible to find out, for instance, the cerebral areas involved. According to transhumanists, we will be able to find the solution and modelize it by means of a digital calculator. Therefore, De Moore "will cross paths" with biology around 2025. It will then be possible to create a brain in a "box" which should be, theoretically, as capable as a human being. According to these scientists, we will be just in time to upload the "minds" of the aging Baby Boomers who will have witnessed the advent of all the new technologies. This would be true if synapses were the only thing that mattered. Our brain is becoming more and more complex as we dig deeper, thus delaying “the optimists’ delight”. New imaging techniques allow us to implement neuron developments in real time. We should therefore be able to find out whether we have to model our brain in more detail or whether all these interactions can be modeled much more simply.

12.5.4. *Future man–machine interfaces*

Man–machine interfaces were conceived in the 1960s. The founders of the man–machine interaction built the first interfaces using the desktop metaphor as inspiration. Thus, in 1963, the Sketchpad, the first graphic interface, was born, followed by the mouse in 1964. We will have to wait until 1981 for the appearance of the Xerox Star, the first machine to fully suggest the desktop metaphor. We should point out that there are no metaphors specific to the health industry, unlike the desktop one used for the Office Automation System (OAS).

Currently, we are in a “touch” phase, defined by interactions which depend on hand movements on a screen. At the moment, the majority of actors are so accustomed to it that they perform actions without even realizing it.

Name of function	Action performed
Flick	Scroll
Pinch	Zoom
Swipe	Change of page

Table 12.1. *Kinds of user-friendliness*

The issue concerning future interfaces is at once simple to describe and complex to turn into a reality. It is a matter of grasping human actions. The trickiest part consists of turning complex and variable interactions into simple and natural ones. Different solutions are being considered: motion capture (Microsoft Kinect), smart glasses (Google), interactive table, the already outmoded vocal recognition, etc.

One of the current tenets is that the development of these innovative interfaces inevitably entails the user's relocation at the center of the design process. It is a matter of predicting both physical and mental behaviors. This is why we need to set up "living labs" where applications can be tested on a full scale. Problems are mainly performance related. First, interfaces have to adapt to physical, cognitive and social abilities. Then, the potential of human communication has to be used as much as possible.

These new interfaces try to make use of the brain immediately and build a dialogue as human as possible. In both cases, this leads to a greater dematerialization of the interaction.

Brain Computer Interactions (BCI) translate cerebral electrical activity into computer inputs. Therefore, giving orders by means of our brain is by now becoming a reality. This development opens up a world of significant possibilities for disabled people. Some researchers, however, consider it the beginning of "telepathic" applications.

The second innovation consists of the notion of autonomous virtual agents (AVA), i.e. software that can interact, verbally and non-verbally, with a user. AVAs can even replicate facial expressions, intonation and reproduce elements of human communication. Then, this kind of solution is brought to life via virtual avatars on a screen or through humanoid robots.

12.5.5. *Immersive augmented reality*

We still have not reached the stage of immersive augmented reality and intelligent software, but technology is getting closer and closer to it. Currently, smart glasses like those recently proposed by some Japanese companies and Google, can be used as an actual superimposed screen for televisions and computers. A firm called Lumus has launched a pair of 3D-display glasses[1]. Anyone can experience different augmented realities. Artificial intelligence, or something similar, is liable to lead to surprising results. Current efforts have managed to find solutions for "the visual cortex", "spatial proprioceptive representation", "the auditory cortex" and the "pleasure center". The programs that transform these "forays" have to allow us to develop software with the potential to travel better than a trillion neurons.

1 The website is available at: http://www.lumus-optical.com/.

12.5.6. *Seeing beyond the visible*

Recent progress in terms of physical lenses (whether made of glass or not), optical fibers and sensors (especially infra-red) now allow us to see beyond the visible. Related applications are being developed in the health industry.

Despite the exceptional performances of our sight, we can only detect a small part of the light spectrum, namely the one with visible wavelengths, which ranges from blue to red. Our natural sensor no longer works if we move past the red part of the spectrum into the domain of heat, which is an important element in the medical field. Therefore, it is possible to conceive a new way of diagnosing. Our visual perception of the world of colors relies on the exceptional optical performances of our sight. A human retina contains around 130 million cells sensitive to light called photoreceptors. Among these, there are 125 rods and 5 million cones. However, our natural sensor can only reveal to us a very small part of the light spectrum, i.e. the one with visible wavelengths, which ranges from blue to red. Rods are slow photoreceptors but they are extremely sensitive to light. They are very useful for night vision and provide us with a black-and-white kind of vision. Cones allow us to perceive colors. There are three kinds of them, depending on whether they react to red, green or blue. The ability to see past the red part of the spectrum and into the infrared field is interesting, since this is the dimension of heat and molecular vibrations. New systems can observe molecular vibrations and give us reason to hope we will be able to analyze some of the metabolisms and systems upon which human life relies. For instance, we will be able to observe the respiratory system in more detail.

Thanks to recent progress in terms of lenses, optical fibers and sensors, especially the infrared ones, it is possible "to see beyond the visible". In the health industry, related applications are being developed. The first of these consists of the development of night vision systems, which were first used by soldiers but can also be useful for people whose night vision has been impaired by rod-related diseases such as retinitis pigmentosa. This disease affects 1 in every 5,000, which corresponds to around 40,000 people in France. Thus, the discovery of new infrared glasses has allowed us to develop a night vision device for medical care teams working at night.

The development of sensors of molecular biological signatures allows us to devise test applications for the identification of metabolic anomalies.

However, these solutions can also be applied to the mental health field: these exotic glasses will let us see reality in a different light, namely through rose-tinted spectacles.

12.5.7. *Sensing emotions and neurosciences*

The study of emotions was first used for marketing. The problem was the following: why do consumers prefer the taste of product A in blind experiments but choose product B once they can see the labels? The solution was found by analyzing the cerebral activity of the "unaware" and conscious users of A and B. Encephalography and fMRI reveal large differences in the activated areas. Other examples, such as the areas involved in language learning, have also been studied. Thanks to functional neuroimaging, it is currently possible to better our understanding of the functions of different areas of the brain. This way of acquiring information goes beyond the mere control system we described when talking about Brain Computer Interface approaches.

Scientists have therefore conceived the possibility of decrypting brain processes, hoping to model the decision-making process. Neuroscientific studies require a significant amount of time and the availability of costly equipment. Moreover, data interpretation remains complex. The unpredictability of human behavior combines with the interactions with the environment, which makes it tricky to separate environmental elements from behavioral aspects. Besides, in several countries certain laws limit the use of MRI and CT scanning. For instance, uses for commercial purposes have been prohibited in France since July 2011. These regulations are running the risk of being changed.

The latest experiments on decision taking mechanisms showed that our brain anticipates action. This prediction takes place before we can even realize it, but it leaves our conscience a period of time to oppose it. This situation led some researchers to claim that we have no free will and that our freedom is only limited to opposition.

12.6. Online data and "big data"

Online data also constitutes a kind of progress.

Administrative documents, scientific articles, radiological and CT scan images are currently stored digitally. However, we should also add some types of digital sources in their own right: emails, Internet searches, activity on social networks, etc. Let us consider some examples. Facebook pumps 500 terabytes of new data into its system per day on its own. The EMC Corporation announced in 2013 that we needed no more than ten minutes to produce 5 terabytes. Ericcson predicts that there will be 50 billion objects (access badges, cars, fridges, smart meters, sensors of every kind, etc.) connected to the Internet globally before 2020, against the dozen billion objects connected nowadays. Out of these, around 10 billion have the potential to generate data for health purposes. Open data, the movement demanding that administrations make public data freely accessible, is also a new source and promoter of the latest developments in the United States, and constitutes a catalyst for "big data".

12.6.1. *Volumes produced*

Thus, giga-, tera- and even petabytes of data are being produced by hospitals as much as GPs. Every expert admits that the volume of data produced doubles every two years. The estimated figures are currently in the range of exabytes, i.e. billions of gigabytes, and even zettabytes (thousands of billions of gigabytes). This data will combine with "big data". Big data affects everyone, whether individuals or structures, like administrations and firms, and, in our case, the healthcare system.

12.6.2. *The predominance of the digital system*

Our memory and our ways of storing information on paper have inexorably migrated toward a digital dimension. Besides the problem of future use and storage costs, we should wonder about what future generations will be able to use. We should become aware of the fragility of this immaterial heritage, since the human lifespan, which in Western countries exceeds 80 years, is much longer than that of storage technologies.

Digital information has a very simple feature: it replicates endlessly without making mistakes. On the other hand, no format known to man can currently ensure that information will be kept for more than a few years.

12.6.3. *The limitations of the digital system*

The main problems are quite widespread:

– the obsolescence of materials: what could we do today with an 8-inch floppy disk?

– the obsolescence of software: how can we re-read the “multiplan” charts of a clinical research?

– natural aging of formats: how can we re-read a damaged CD-ROM or a magnetic tape that is no longer working?

– hard disk errors or bugs: how can we find again the data that was inexorably deleted from such a format?

The myth of written information is that it was “set in stone”. This is no longer the case for these formats and their software. Current data requires a perpetual process of copying out and management. In other words, each digital object left to its own devices runs the risk of getting inexorably lost.

The short life expectancy of formats with regard to the lifespan of humans constitutes a real problem for applications in the health industry. According to some, medical data must be condemned to endlessly shift toward new formats. For others, the “cloud” is the solution that would allow the management of data networks. The volatility of data created digitally is a reality and its disappearance a constant feature.

Digital archiving, introduced as a necessary element from the very beginning, is a new issue that comes into play for the preservation of the memory of the 20th and 21st Century. We have to point out that this is not always the case. We should also remark that several scientific journals, some of them medical, have disappeared together with conference organizers, and sometimes even within great academic associations, the names of which I would never dare to mention here. Currently, the actors involved in the information field have gone even further. At first, let us point out that this development is made possible by the steep drop in storage costs, which depend on the famous Moore’s law but speed up the regeneration, and consequently the obsolescence, of the media.

12.6.4. *Volume, variety, velocity, value, veracity and infobesity*

Big data will be "the next frontier for innovation, competition and productivity", according to a consulting firm called McKinsey, which wrote a report on this topic in 2011. One of its main features, apart from volume, is the variety of data formats (texts, videos, audio, transactions, sensor data, etc.).

This data can actually barely be processed, stored and analyzed with IT traditional tools. Database management and analytic decision-making have become obsolete. Even worse, my favorite classic visualization methods, such as graphics, can no longer present this information in an intelligible fashion.

A new term has emerged: "infobesity". In order to understand this type of IT obesity, assessment tools based on the five Vs (volume, variety, velocity, value, veracity) have been introduced.

12.6.5. *Solutions to infobesity*

One of the solutions consists of the weeding operation, carried out by paper-age archivists, now called curation. This weeding consists of getting rid of the least read books of a public library and of curating the data that seems less useful. The main problem derives from the use of these concepts in relation to medical and health data.

Wikipedia provides a definition of curation:

"Content curation is a process consisting in selecting, editing and sharing the digital content most pertinent to a query or given topic. Curation is used and claimed by websites that wish to improve the visibility and intelligibility of some content (such as texts, documents, images, videos, audio files, etc.) that they deem useful for users and that, when shared, can help them or concern them.

Content curation falls within the circle of influence of the Semantic Web, a better organized ecosystem that allows machines to process users' queries more intelligently and to display more pertinent results pages".

Some do this more or less deliberately, others leave it to chance whereas the "wisest" claim to establish certain rules.

12.6.6. *From documents to contents*

The beginning of the Office Automation System (OAS) was defined in the 1980s by the desktop and the document metaphors. Actors stored documents "analogous to" papers. However, the 2000s saw the advent of content management solutions. Therefore, health, like other fields, shifted from document management to the management of the information included in the document. Personal medical files, among other things, embody this development. Little by little, IT succeeded in managing data. Then, it was possible to analyze this information thanks to new technologies called "big data".

Thus, we get to another problem linked to this avalanche of data. This data, once combined, filtered and analyzed, becomes a formidable tool in the hands of the actors involved, hence this veritable frenzy. This hope immediately clashes with the question concerning the respect for personal data. This situation could be summed up as respect for privacy versus big companies. Therefore, insurance agents can better assess the risk if the patient's medical data can be accessed.

Another problem pertains to the quality of data, which is crucial in the health industry and was pointed out by a firm called IDC. Thus, 23% of digital data could be useful for "big data" if it were well marked ("tagged") and analyzed, whereas only 3% actually is. We can experience firsthand the weak performance of the "big data" system. A lot of content is worthless. At a first stage, this justifies the concept of curation.

12.6.7. *Effective use*

According to some reports, the effective use of data in the health industry could decrease costs by 15–20%, reduce the incidence of medical errors and increase our lifespan. However, no serious research has verified these figures. The health industry is one of the most affected areas. Thus, big data programs are set up in such fields as genomics, proteomics and medical imaging. In the first two cases, the matter boils down to processing more quickly huge data sets for high throughput sequencing. As for medical imaging, the question pivots around the availability of tools for quick and effective detection. Other contributions in the health industry are expected. They should translate into hugely improved effectiveness in terms of detection of medical errors, predictive medicine and optimization of operational care protocols. The analysis of the performances of medications (medical service rendered) could be effectively carried out during their clinical trial before their launch on the market, and then afterwards to assess their effectiveness and prevent the appearance of adverse effects. Geolocation data will allow us to analyze the development of viral or bacterial attacks.

12.6.8. *Professions*

Certain professions are associated with this progress:

– data scientist, i.e. an expert in statistical methods. He is competent in interpreting and cross-referencing data.

– chief data officer (CDO), responsible for the quality of data.

– engineer and developer of applications adjusted to "big data".

12.6.9. *New software frameworks*

Hadoop is an open-source software framework written in Java and developed by an engineer working for Yahoo, now being managed by Apache Software Foundation. This platform works like an operating system that manages computer clusters. Therefore, it allows us to process very large "unstructured" multiformat data sets. Hadoop has been adopted by nearly all the big data giants (Yahoo, Facebook, Amazon, IBM, HP, SAP, etc.). The definitive version was launched in 2011. Alternative solutions are being developed (Cassandra, Isilon, etc.). New "massively parallel" computing methods, such as MapReduce, have been proposed by Google.

12.6.10. *The patient's data*

In order to allow the simultaneous implementation of medical research, personalized medicine and effective treatments, patients' medical data must necessarily be recorded.

The successful development of e-health in Nordic countries derives from the separation between medical and personal data. Thus, the Electronic Patient Record (EPR) consists of data gathered by healthcare professionals and meant for other healthcare professionals. On the other hand, the Personal Health Record (PHR) includes the health data belonging to an individual, whether he is a patient or not. There is in fact an interaction between the two records; however, PHRs can be collected regardless of medical treatment, thus creating the basis for a healthy patient.

Therefore, the EPR is a tool used by doctors, whereas the PHR is one of the means that individuals use to become *santacteurs*. Currently, this data often begins to be gathered when the individual becomes a patient. It is actually interesting to store medical data over a lifetime, hence the question concerning *santacteurs* and health data.

12.6.11. *Opt-in, anonymity and limitations*

The massive use of health data makes the question of privacy more prominent. According to the French National Commission on Informatics and Liberty and the European Commission, we should remain vigilant to the ways this data is being used and make sure that certain rules are respected. The main one consists of the preliminary consent of a patient, whether he has become a patient or not. The Commission has drawn up a directive on personal data, which may be "a name, a photo, an email address, bank details, posts on social networking websites, medical information, or a computer IP address". Therefore, any failure to comply with these rules is subject to fines of up to 1 million euros, or 2% of the global yearly revenue of a company.

We are thus faced with one of the limitations of "big data", namely the existence of data itself. Let us imagine that all the carriers of a virus have availed themselves of the opt-in clause. As a result, the disease could no longer be detected.

12.6.12. *Unfeasible exhaustiveness*

Thus, we are part of an era characterized by unfeasible exhaustiveness due to opt-in and curation. Therefore, we will be compelled to anticipate the doctors' need for data in order to ensure, in a first phase, the continued existence of digital information. The second step will consist of developing technologies adapted to the health industry. These approaches will ensure a long-term form of archiving that will even manage to outlive every individual. They will also allow us to "piece together" the missing data. Big data is, in any case, one of the approaches that will be used by the industrialization of health.

12.7. Robots and robotics

Robotics is one of the technologies that is becoming more and more widespread in the medical field, whether to be used as a tool in surgical operations or to assist patients, etc.

12.7.1. *Robotics*

In a few years, "our" workers' main competitors will not be immigrants or exotic countries, but robots. Repetitive tasks are meant to be automated. Jobs are being redeployed in relation to the conception of products and the services they entail. Workforce will evolve into brainforce. Thus robots are replacing man's body and they will soon do the same with his mind. Human beings have always tried to reduce both physical and intellectual efforts and to find solutions that would allow them to avoid performing repetitive tasks.

A robot is a mere technical element of a sociotechnical organization to which humans delegate, and theoretically yield, the power to make decisions within the organization itself, which entails the power to perform. The notion of control raises one question. The problem posed by robotics is analogous to the one involved in the introduction of paper and pencil, then machines-tools and more recently information systems. There is, however, a slight difference: robots act on man himself.

After the 20th Century's IT revolution, our era seems characterized by an actual revolution of sensors and robotics. Robotics was born in the 1960s in research laboratories. The MIT and Stanford, in the United States, are often mentioned but we should not forget about the projects carried out in Europe, such as the ARA program in France.

Robots had already been evoked in the writings of the great Romantic authors of the 19th Century. They had already been a reality well before they were actually invented. Hoffmann, Villiers de L'Isle-Adam, Mérimée, Balzac, Poe and others introduced the notion of mechanical humanoid forms who could take the shape of mechanical men, living statues, Golems, monsters like Frankenstein, fatal gynoids or universal robots. Thea von Harbou's 1926 sci-fi novel was made into a movie by Fritz Lang in 1927, which allegedly represents the first time robots appeared on screens.

These texts turned robots into mythical creatures that arouse fascination as much as fears, which are factors that certainly contributed to the current notion of robots, hence the clash between two "schools of thought" on the design of future robots. Some imagine an uber-specialized robot that will be designed according to its function, whereas others conceive a humanoid robot at once easily programmable and very versatile. As for the latter notion, the idea of a robot replacing man seems currently unfeasible. However, our robots are becoming more and more efficient.

At first, robots were mere workers dealing with items or industrial elements. Afterwards, they evolved and can now move both themselves and other objects, cut and assemble, weld and paint, etc. Currently, trains and subways are evolving and function more and more on "autopilot". All these automated systems are presently being developed. Fitted with several visuals, sounds and inertial sensors, they acquire new sensorimotor skills. They can therefore explore unknown environments while interacting with humans, at least at a theoretical level. They often give us the impression that something else will carry out the tasks performed by man. Therefore, they raise several economic problems and, first of all, the question of the employees' future and redeployment. Another problem pertains to the capital costs involved in the construction of these substitutes.

Industrial robotics has therefore been developing for quite a long time and has even become predominant in hospital operating rooms. The first industrial robot was born 50 years ago as part of a GM production line. Large manufacturers regularly announce they are fitting out their factories with millions of robots. Is this a new era in which robots can replace man?

More recently, there was a real breakthrough in terms of service robots, which took place in two main ways. The first kinds of robots tried to play with man, the latter being in control. The second type of robots attempted to render some services. At the beginning of the 2000s, the famous dog robot called Aibo broke ground. It tested the cohabitation potential between men and robots. The production came to a halt in 2006 and no real reasons were given. It was then followed by humanoid robots. Nao, a small humanoid robot frequently used by researchers, teachers and developers, could perhaps become a companion robot. Presently, there are about 20 other prototypes.

A very common kind of domestic robot, robot vacuum cleaners can nimbly make their way between pieces of furniture and go back to their charger on their own.

The main challenge faced by current robotics has to do with men's ability to master their robots and vice versa. Humanoid robots require the institution of new types of coexistence.

Domestic robots are no longer relegated to separate dimensions. The second challenge consists of enabling these tools to evolve in an environment conceived by men. Men and machines have to learn to co-operate in a normal setting. Certain kinds of relationships have to be established and the necessary decision-taking processes have to be defined.

This challenge refers to the problem posed by the notion of intelligence. If these machines become more and more intelligent, we can regard this situation as either a mere illusion or a solution, depending on whether we adopt a transhumanist perspective or not.

The main questions are:

– which tasks should be automated?

– to what extent does man accept machines?

– what kind of equilibrium should there be between automated systems and human activities?

12.8. Selection technologies

12.8.1. *Selection*

Choosing between individuals "liable" to fall sick and other people will become a reality, but it will also raise ethical problems. We have already tackled in this book the case of women carrying the BRCA gene.

The press regularly spreads news about scientists that have allegedly collected DNA samples belonging to the smartest people in the world. Thus, they have allegedly sequenced their whole genomes in order to attempt to identify the alleles (genomic sequences) responsible for human intelligence. Others propose embryo screenings that will allow parents to choose their "best" zygote and to potentially increase the intelligence of each generation by 5–15 IQ points. This is basically a variation of the eugenic approach described in certain novels. Couples would fertilize about a hundred zygotes (embryos) and then science would analyze them. They would then choose one of them, go through pregnancy and give birth to a completely natural baby.

For decades, "phage therapy" was the domain of medicine. The idea, explored mainly by scientists, consisted of the hope of finding viruses that would preferentially infect and kill those kinds of microbes that are parasites for humans. Another solution consists of creating viruses that only attack human cancer cells. An oncolytic virus is a virus that infects and kills cancer cells while leaving healthy tissue unaffected. The notion of oncolytic virus has long been used by sci-fi and was first introduced to the public in Jack Williamson's novel *Dragon's Island* (1951). The idea of this development has recently reappeared and some clinical trials had encouraging results. Amgen acquired an oncolytic virus company called BioVex for a billion dollars in January 2011. More recent news seems to suggest that a phage will soon attack melanomas in the patients affected.

12.8.2. *Screening tests, companion diagnostics*

The potential of current tests goes well beyond mere screening. Companion diagnostic or prognostic markers are the two terms used most frequently. Presently, we have at our disposal around 20 companion diagnostics for colon, stomach, lung and breast cancers, and for some kinds of leukemia and melanoma.

Roche laboratories are one of the promoters of this solution and offer a predictive test together with the prescription of Xalori (Pfizer) in cases of lung cancer. This product binds only to cancer cells and works exclusively on those patients that present an anomaly targeted by the treatment.

12.9. Health and technological flops

Implementations in the health industry are characterized, among other things, by the attempt to re-use applications that were “technological flops”.

12.9.1. *The multitude of examples*

Health often involves an application put forth and presented as revolutionary. At the end of the 1970s and in the early 1980s, videotelephony seemed like the most significant application of the century. Telemedicine was born. The Biarritz experiment had even verified this hypothesis. Some years later, a large manufacturer of telecommunications material justified the birth of 3G technology with a few clips, one of which showed an emergency intervention to the site of a car accident in which first-aid workers communicated with doctors via videophone. More than 30 years later, we have to ask this question: why has videophony never managed to establish itself? The first kinds of fridges were linked to cinema, and then TV commercials that praised a future smart fridge able to detect the products stored and help people come up with a menu. This kind of fridge suddenly reappears together with the hope that elderly or dependent people can remain in their homes. The smart health house is a set of these kinds of products. Why doesn’t anyone then, 60 years later, own this product which has so often been praised for its qualities?

12.9.2. *The problem of diffusion*

A first answer consists of the hypothesis that it is quite tricky for those innovations that try to revolutionize our daily lives to become widespread. However, this turns out to be untrue when we take into account communication technologies.

Let us look at the facts. First of all, several prototypes never manage to make their way out of laboratories and, consequently, we never get to hear anything about them. Other products reach the status of dead-end media or “one shot”: they are given media coverage and showcased just to be soon afterwards abandoned due to lack of use. Sometimes, the future customer does not find them interesting, whereas in other cases it is a matter of technologies too advanced for their environment. In these two cases, these technologies can be reproposed afterwards. Telemedicine is supposedly the future of videophony.

Exceedingly futuristic concepts and products are often means through which the future customer is made to react to these wishes. This kind of demonstration has been used for products such as cars but when is it used in the health industry? The matter boils down, in this domain, to the problem posed by a demonstration coming before the proof.

The causes of technological flops are at once varied and complex.

12.9.3. *Some theoretical notions to get over flops in the health industry*

Some theoretical concepts have been proposed.

The first one has to do with the technological tradition. The ideas of innovative projects are often the subject of several variations and revisions. None of them reaches the market. This situation lasts until one of the products following the "technological tradition" becomes successful thanks to users who start making use of it. In these precise circumstances, success appears after a series of failures.

The second theoretical concept pertains to misapplications. The products invented are therefore used in some ways that were not initially considered.

The third concept proposed has to do with error analysis. This solution, which makes some advisors happy, is not simple. It is often based on reviewing past literature and the collective examination of former failures. This study of experiments ended in commercial failures has several advantages. At first it allows us to downplay the failures. It makes us acknowledge a right to error, which is the basis for experiments and the mainstay of innovation. Finally, it enables us to improve the methods used to conceive solutions.

Conclusion

We have only one certainty: the industrialization of medicine will not lead to enhanced men, androids, bionic men or cyborgs. It will lead to a new and different kind of man. Gerontechnology, the robolution, all kinds of anthropotechnics and transhumanism will be ways in which this theory is created. However, none of them will be the only approach used.

The industrialization of health will not be limited to engineering science, which cannot deal with the complexity of health. This point is certainly the core of the imminent scientific revolution, which will require trust and awareness in terms of identity. We still need to investigate the power relations (biopower) that will be established.

Until now, the only actual complex systems have been nature and the evolution process. However, man has now been able to alter his environment and has affected nature by means of medical progress. Therefore, infant mortality has decreased and contraception has affected the size of families. It is not a question to turn down this situation. We only have to realize that man is no longer one of the elements of nature, but an adaptable complex system joined by another similar system, i.e. health. Man relates to nature and is equipping himself with new tools that the industrialization of health will enable him to build.

Bibliography

[ACA 13] ACADEMIE NATIONALE DE PHARMACIE, Medicinal products: stock-outs and supply disruption, report, 20 March, 2013.

[AKR 87] AKRICH M., "Comment décrire les objets techniques", *Techniques et culture*, no. 9, 1987.

[BAR 91] BARNEY J., "Firm resources and sustained competitive advantage", *Journal of Management*, vol. 17, no. 1, pp. 99–120, 1991.

[BEN 06] BENDERSON B., *Transhumain*, Payot, Paris, 2006.

[BER 15] BERANGER J., *Medical Information Systems Ethics*, ISTE, London and John Wiley & Sons, New York, 2015.

[BON 10] BONNELL B., *Viva la Robolution. Une Nouvelle Étape pour l'Humanité*, JC Lattès, 2010.

[BOU 79] BOURDIEU P., *Distinction: A Social Critique of the Judgment of Taste*, Cambridge University Press, Cambridge, MA, 1979.

[BOU 81] BOURDIEU P., *Questions de Sociologie*, Editions de Minuit, Paris, 1981.

[BOU 13] BOULIER D., HECKER A., Medialab Science Po, available at: http://www.medialab.sciences-po.fr/fr/projects/habitele-identites-numeriques-portables, 2013.

[BRU 08] BRUNS A., *Blogs, Wikipedia, Second Life and Beyond: From Production to Produsage*, Peter Lang, New York, 2008.

[CAM 08] CAMUS R., *La Grande Déculturation*, Fayard, Paris, 2008.

[CAS 75] CASTORIADIS C., *Institution Imaginaire de la Société*, Seuil, Paris, 1975.

[CAS 12] CASAGRANDE A., *Ce que la maltraitance nous enseigne, Difficile bientraitance*, Collection: Action Sociale, Gazette Santé Social, p. 224, 2012.

[CHA 09] CHARDEL P.-A., *Technologies de Contrôle dans la Mondialisation: Enjeux Politiques, Éthiques et Esthétique*, Kimé, 2009.

[COM 98] COMTE A., *Course in Philosophy*, Éditions Hermann, 1998.

[DEG 13] DEGOS L., *Éloge de l'Erreur*, Le Pommier, Paris, 2013.

[DUR 07] DURKEHIM E., *The Division of Labor in Society*, PUF, Paris, 2007.

[ELL 12] ELLUL J., *Le Système Technicien*, Calmann-Lévy, Le Cherche-midi, 3rd ed., Paris, 2012.

[FER 06] FERRARI M., *T'Es où? Ontologie du téléphone mobile*, Albin Michel, 2006.

[FLI 10] FLICHY P., *Le Sacre de l'Amateur*, République des Idées Collection, Seuil, Paris, 2010.

[FOM 96] FOMBRUN C.J., *Reputation: Realizing Value from the Corporate Image*, Harvard Business School Press, Harvard, 1996.

[GAL 03] GALIBERT O., Les communautés virtuelles: entre marchandisation, don et éthique de la discussion, SIC Thesis, Stendhal University, 2003.

[GEN 99] GENSOLLEN M., "La création de valeur sur internet", *Réseaux*, vol. 17, no. 97, p. 15, 1999.

[GEN 04] GENEL K., "Le biopouvoir chez Foucault et Agamben", *Methodos*, available at: https://methodos.revues.org/131, 2004.

[GHA 09] GHAFERI A.A., BRIKMEYER J-D., DIMICK J.B., "Variation in hospital mortality associated with impatient surgery", *New England Journal of Medicine*, vol. 362, pp. 1368–1375, 2009.

[GIR 61] GIRARD R., *Romantic Lie and Romanesque Truth*, *Fayard/Pluriel*, 1961.

[GOD 97] GODELIER M., *The Enigma of the Gift*, Fayard, Paris, 1997.

[GOF 06] GOFFETTE J., *Naissance de l'Anthropotechnie, De la Médecine au Modelage de l'Humain*, Vrin, 2006.

[GRI 08] GRINEVALD J., *La biosphère de l'anthropocène: climat et pétrole, la double menace?*, Geog édition, Geneva, 2008.

[HEN 78] HENDERSON V., NITE G., *Principles and Practice of Nursing*, Macmillan, New York, 1978.

[KUR 12] KURZWEIL R., *How to Create a Mind, the Secret of Human Thought Revealed*, Viking Adult Editor, 2012.

[LED 14] LE DEVEDEC N., *La société de l'amélioration. La perfectibilité humaine, des lumières au transhumanisme*, Liber, Montreal, 2014.

[LEH 07] LEHIKOINEN J., AALTONEN A., HUUSKONEN P. *et al.*, *Personal Content Experience: Managing Digital Life in the Mobile*, Wiley, London, 2007.

[LEV 13] LEVY LEBLOND J.M., *Le Grand Écart, la Science entre Technique et Culture*, Manucius, Paris, 2013.

[MAI 04] De MAILLARD J., VEYRIER J.C., *Le Rapport Censuré: Critique non autorisée d'un monde déréglé*, Flammarion, 2004.

[MAL 94] MALLEIN P, TOUSSAINT Y., "L'intégration sociale des TIC: une sociologie des usages", *Technologie de l'information et société*, vol. 6, no. 4, pp. 315–335, 1994.

[MAT 10] MATTELART A., *The Globalization of Surveillance*, Polity Press, 2010.

[NEG 05] NEGRI A., HARDT M., *Multitude: War and Democracy in the Age of Empire*, Penguin, London, 2005.

[OMR 71] OMRAN A., "The epidemiological transition: a theory of the epidemiology of population change", *The Milbank Quarterly*, vol. 83, no. 4, pp. 731–57, 1971.

[PEN 59] PENROSE E., *The Theory of the Growth of the Firm*, John Wiley and Sons, Hoboken, NJ, 1959.

[PRA 90] PRALAHAD C.K., HAMEL G., "The core competencies of the corporation", *Harvard Business Review*, vol. 68, no. 3, pp. 79–93, 1990.

[RAW 05] RAWLS J., *A Theory of Justice*, Harvard University Press, Cambridge, MA, 2005.

[RIF 83] RIFKIN J., *Algeny*, Penguin Books, New York, 1983.

[RIN 98] RINDOVA A., FOMBRUN C.J., "Constructing competitive advantage: the role of firm-constituent interactions", *Strategic Management Journal*, vol. 20, pp. 691–710, 1998.

[ROG 95] ROGERS E., *Diffusion of Innovation*, 4th ed., The Free Press, New York, 1995.

[ROS 06] DE ROSNAU J., REVELLI C., *La Révolte Du Pronétoriat-Des Mass media aux media des masses*, Fayard, Paris, 2006.

[SAP 04] SAPIRO G., "Une liberté contrainte. La formation de la théorie de l'habitus", *in Pierre Bourdieu, Sociologue*, Fayard, p. 69, 2004

[SCH 43] SCHUMPETER J., *Capitalism, Socialism and Democracy*, Payot, Paris, 1943.

[SEA 10] SEABRIGHT P., *The Company of strangers: A Natural History of Economic Life*, Princeton University Press, 2010.

[SEN 73] SEN A., *On Economic Equality*, Norton, New York, 1973.

[SLA 97] SLATER D., *Consumer Culture and Modernity*, Polity Press, Cambridge, UK, 1997.

[SOR 01] SORIANO P., FINKIELKRAUT A., *Internet, l'Inquiétante Extase*, Éditions Milles et une Nuits, Paris, 2001.

[TON 87] TONNIES F., *Community and Society (Gemeinschaft und Gesellschaft)*, 1887.

[WER 99] WERNER A., WERNER H., GOETSCHEL N., *Les Épidémies, un Sursis Permanent*, Atlande, Paris, 1999.

[WOO 91] WOOLGAR S., *A Sociology of Monster, Essays on Power, Technology and Domination*, Rouledge, London, 1991.

Index

A, B, C

D, E, F

G, H, I

J, K, L, M

N, O, P, Q

R, S, T

U, W, Z

Printed in the United States
By Bookmasters